移动互联网开发技术丛书

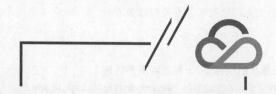

微信小程序云开发
超详细实战攻略

姜丽希 厉旭杰 主编

清华大学出版社

北京

内 容 简 介

本书通过通俗易懂的语言、丰富实用的案例，全面系统地介绍了微信小程序云开发所涉及的知识。

全书包括两大部分。第一部分主要介绍微信小程序云开发的基础知识，包括第1～5章，讲述微信小程序与云开发相关基础知识，云数据库、云存储和云函数等功能。第二部分为微信小程序云开发实战，包括第6～16章，精选了多个简单易学的实例，通过一个个案例把微信小程序云开发技术运用到实际的问题当中，最后两章涉及通过Web端访问云开发资源，并提供Web API接口。

本书适合微信小程序云开发的初学者（特别是在校学生）、微信开发者和前端开发爱好者阅读，也可以作为学校的教学用书、自学的入门读物和开发过程的参考书。

本书封面贴有清华大学出版社防伪标签，无标签者不得销售。

版权所有，侵权必究。举报: 010-62782989, beiqinquan@tup.tsinghua.edu.cn。

图书在版编目(CIP)数据

微信小程序云开发超详细实战攻略/姜丽希，厉旭杰主编.—北京：清华大学出版社，2021.1
(移动互联网开发技术丛书)
ISBN 978-7-302-56399-0

Ⅰ.①微… Ⅱ.①姜…②厉… Ⅲ.①移动终端-应用程序-程序设计 Ⅳ.①TN929.53

中国版本图书馆CIP数据核字(2020)第170449号

责任编辑：陈景辉　张爱华
封面设计：刘　键
责任校对：白　蕾
责任印制：杨　艳

出版发行：清华大学出版社
网　　址：http://www.tup.com.cn, http://www.wqbook.com
地　　址：北京清华大学学研大厦A座　　　　邮　编：100084
社 总 机：010-62770175　　　　　　　　　　邮　购：010-83470235
投稿与读者服务：010-62776969, c-service@tup.tsinghua.edu.cn
质量反馈：010-62772015, zhiliang@tup.tsinghua.edu.cn
课件下载：http://www.tup.com.cn, 010-83470236

印 刷 者：北京富博印刷有限公司
装 订 者：北京市密云县京文制本装订厂
经　　销：全国新华书店
开　　本：185mm×260mm　　印　张：24.25　　字　数：588千字
版　　次：2021年1月第1版　　　　　　　　　印　次：2021年1月第1次印刷
印　　数：1～2000
定　　价：89.90元

产品编号：087908-01

前言
FOREWORD

  传统的微信小程序开发需要考虑服务器的性能、负载均衡、网络安全等一系列运维问题。这些工作非常烦琐而且耗费时间精力，那么为什么我们不将它交给专业运维的人去配置呢？云开发可以很完美地帮我们解决以上问题。云开发是云端一体化的后端云服务，采用Serverless架构，免去了移动应用构建中烦琐的服务器搭建和运维。微信小程序云开发极大地提高了微信小程序的开发效率，开发者只需要关心应用层面的业务逻辑和用户的交互体验。

  全书共16章，分为微信小程序云开发基础和微信小程序云开发实战两部分。第一部分是微信小程序云开发基础，包括第1~5章。其中，第1章是微信小程序云开发介绍，详细讲解什么是云开发、如何注册微信小程序、如何新建云开发项目以及如何对项目进行初始化；此外还介绍微信小程序常见UI组件库和图标库。第2章是微信小程序相关基础知识，详细介绍微信小程序开发中ECMAScript 6语法的使用以及微信小程序框架。第3~5章详细介绍云数据库、云存储和云函数。第二部分为微信小程序云开发实战，包括第6~16章。其中第6~13章引入了8个实用的案例，包括新闻微信小程序、投票微信小程序、通讯录微信小程序、报修微信小程序、网上书城微信小程序、团购类微信小程序、会议室预约微信小程序和AI+微信小程序。第14章详细介绍在微信小程序中如何使用ECharts，第15章介绍如何在Web后端通过HTTP API访问云开发资源，第16章以目前最流行的Vue.js前端技术为例，详细介绍如何通过SDK访问云开发资源。

  本书特点

  （1）知识全面，循序渐进。

  书中详细介绍微信小程序云开发技术及其相关知识，通过具体实战案例，运用腾讯云短信服务、百度AI识别、数据图表ECharts等技术，帮助读者深化学习。

  （2）案例实用性强，可直接上线发布。

  书中的案例均以云开发技术来实现，每个案例都可以独立运行，并可以直接上线发布。案例通俗易懂，实用性强，所有案例均来自已经上线的微信小程序项目。

  （3）学习门槛低。

  本书最大的特色在于读者仅需掌握基本的JavaScript知识，即可通过学习本书快速地开发出微信小程序项目，而无须额外地学习后端技术，只需要关注页面业务逻辑，即可快速开发出实用、美观的微信小程序项目。

配套资源

为便于教学,本书配有源代码、数据集、案例素材。

获取源代码、数据集、案例素材方式:扫描下方相应的二维码,即可获取。

源代码

数据集

案例素材

读者对象

本书适合微信小程序云开发的初学者(特别是在校学生)、微信开发者和前端开发爱好者阅读,也可以作为学校的教学用书、自学的入门读物和开发过程的参考书。

本书在编写过程中,濮济、林选、王怡婷、潘瑜同学提供了部分的项目案例,在此对他们表示衷心的感谢!

由于编者水平有限,书中难免存在疏漏之处,敬请广大读者提出修改意见,以便本书再版时修订。

编 者

2020 年 11 月

第一部分 微信小程序云开发基础

第1章 微信小程序云开发介绍 3
1.1 什么是云开发 3
1.1.1 传统的微信小程序开发瓶颈 3
1.1.2 什么是微信小程序云开发 3
1.1.3 微信小程序云开发的核心能力 5
1.2 注册微信小程序 6
1.3 新建微信小程序云开发项目 8
1.4 初始化项目 14
1.5 微信小程序 UI 组件库及图标库 16
1.5.1 常用微信小程序 UI 组件库 16
1.5.2 ColorUI 组件库 18
1.5.3 Vant Weapp 组件库 20
1.5.4 iconfont 图标库 23
1.6 微信小程序优化建议 24

第2章 微信小程序相关基础知识 26
2.1 ECMAScript 6 基础知识 26
2.1.1 ECMAScript 6 简介 26
2.1.2 ECMAScript 变量 29
2.1.3 ECMAScript 数组和对象 32
2.1.4 ECMAScript 语句 37
2.1.5 ECMAScript 6 异步操作和 Async 函数 39
2.2 微信小程序框架 43
2.2.1 注册页面的使用 43
2.2.2 页面路由 45
2.2.3 视图层 WXML 47

2.2.4　this.data 和 this.setData 的区别 ·················· 50

第 3 章　云数据库 ························· 54

3.1　云数据库上手 ························· 54
3.2　数据迁移 ··························· 56
3.3　基础概念 ··························· 61
3.4　云数据库 API 列表 ······················· 63
3.5　云数据库操作 ························· 66
  3.5.1　增加记录 ······················· 67
  3.5.2　查询记录 ······················· 67
  3.5.3　更新数据 ······················· 73
  3.5.4　删除数据 ······················· 75
  3.5.5　正则表达式查询 ···················· 76
  3.5.6　查询和更新数组元素和嵌套对象 ············· 80
  3.5.7　数据库操作 data 赋值 ················· 85
  3.5.8　增、删、改、查案例 ·················· 88

第 4 章　云存储 ························· 99

4.1　管理文件 ··························· 99
4.2　存储 API ·························· 100
4.3　存储操作 ·························· 101
4.4　云存储案例 ························· 102

第 5 章　云函数 ························· 110

5.1　云函数发送 HTTP 请求 ···················· 110
5.2　云函数将数据库数据生成 Excel ················· 118
5.3　本地调试 ·························· 123
5.4　定时触发器 ························· 125
5.5　云函数高级用法——TcbRouter ················· 127

第二部分　微信小程序云开发实战

第 6 章　新闻微信小程序 ····················· 137

6.1　授权页面 ·························· 137
6.2　添加新闻页面 ························ 139
6.3　新闻主页 ·························· 146
6.4　新闻详情页 ························· 151

第 7 章　投票微信小程序 158

7.1　授权页面 158
7.2　添加投票页面 160
7.3　投票主页 167
7.4　投票页面 170

第 8 章　通讯录微信小程序 178

8.1　项目主页 178
8.2　通讯录页面 181
8.3　删除人员页面 192
8.4　添加人员页面 197

第 9 章　报修微信小程序 202

9.1　腾讯云短信平台 202
9.2　登录页面 207
9.3　项目主页 217
9.4　用户报修页面 220
9.5　管理员管理页面 224
9.6　填写报修单页面 228
9.7　报修单页面 236

第 10 章　网上书城微信小程序 247

第 11 章　团购类微信小程序 248

11.1　项目主页 250
11.2　商品分类页面 255
11.3　商品详情页面 265
11.4　购物车页面 267
11.5　提交订单页面 273
11.6　我的订单页面 276

第 12 章　会议室预约微信小程序 279

12.1　项目主页 279
12.2　会议室预约情况页面 285
12.3　填写预约单页面 291
12.4　我的预约页面 297

第 13 章　AI＋微信小程序 ·········· 301

- 13.1　百度 AI 开放平台 ·········· 301
- 13.2　百度 AI 开放平台接口测试 ·········· 306
- 13.3　项目主页 ·········· 309
- 13.4　车牌识别页面 ·········· 311
- 13.5　通用物体识别页面 ·········· 317

第 14 章　在微信小程序中使用 ECharts ·········· 323

- 14.1　在项目中引入 ECharts ·········· 323
- 14.2　项目主页 ·········· 326
- 14.3　柱状图页面 ·········· 328
- 14.4　散点图页面 ·········· 331
- 14.5　折线图页面 ·········· 334
- 14.6　多个图表页面 ·········· 336

第 15 章　通过 HTTP API 访问云开发资源 ·········· 341

- 15.1　搭建 Node.js 网站 ·········· 341
- 15.2　访问云平台数据库资源 ·········· 344
  - 15.2.1　取 access_token 值 ·········· 344
  - 15.2.2　数据库导入 ·········· 347
  - 15.2.3　数据库查询记录 ·········· 349
  - 15.2.4　数据库插入记录 ·········· 352
  - 15.2.5　数据库更新记录 ·········· 353
  - 15.2.6　数据库删除记录 ·········· 355
- 15.3　访问云平台存储资源 ·········· 356
  - 15.3.1　获取文件上传链接 ·········· 356
  - 15.3.2　获取文件下载链接 ·········· 358
  - 15.3.3　删除云存储文件 ·········· 360
- 15.4　触发云函数 ·········· 361

第 16 章　在网页端通过 SDK 访问云开发资源 ·········· 363

- 16.1　创建 Vue.js 项目 ·········· 363
- 16.2　云控制台访问云开发资源 ·········· 366
- 16.3　项目主页 ·········· 368
- 16.4　云数据库操作 ·········· 370
- 16.5　云存储操作 ·········· 375
- 16.6　云函数操作 ·········· 378

第一部分

微信小程序云开发基础

第 1 章

微信小程序云开发介绍

1.1 什么是云开发

1.1.1 传统的微信小程序开发瓶颈

传统的微信小程序开发需要开发人员购买服务器/域名,部署服务器环境,配置 SSL 证书,配置服务器信息。业务逻辑上要使用数据库,实现数据接口。购买、搭建和配置这些内容需要花费不少人力、物力,成本压力大。

以下列出当前开发微信小程序遇到的瓶颈:

(1)需要程序员编写后台代码实现业务逻辑,比如编写最简单的 CRUD(增、删、改、查)需要不少代码;

(2)开发过程中需要对数据库进行操作;要求程序员熟悉 SQL 语句,增加其学习成本;

(3)需要完成会话服务和文件上传保存等工作,需要配置后台服务器,而且安全性不高;

(4)最初的微信小程序是基于 API 来开发的,开发效率较低;

(5)编写和调试用户登录和微信支付的代码十分复杂。

那么,在微信小程序开发中,是否有一种服务,可以释放对后端运维的压力,只需程序员关注好业务逻辑?微信小程序云开发应运而生。

1.1.2 什么是微信小程序云开发

微信小程序云开发是腾讯云和微信团队联合开发的,集成于微信小程序控制台的原生 Serverless 云服务,解决了 Serverless 架构对端的"最后一公里"问题,通过集成端 SDK,配合云开发后台的 API 网关,为开发者提供了一站式后端云服务。云开发支持多种客户端,帮助开发者统一构建和管理资源，免去了开发中服务器的搭建，极大地简化了 URL 配置、鉴权管理等流程，让微信小程序开发者专注于业务逻辑的实现，而无须理解后端逻辑及服务器运维知识，门槛更低，效率更高，只需要一名开发人员就可以完成所有的工作。

Serverless 的中文含义是"无服务器"，但是它真正的含义是开发者再也不用过多考虑服务器的问题，但是并不代表完全去除服务器，而是依靠第三方资源服务器后端计算服务来执行代码，Serverless 架构分为 Backend as a Service（BaaS：后端即服务。它的应用架构是由大量第三方云服务器和 API 组成的，应用中关于服务器的逻辑和状态都由服务提供方来管理）和 Functions as a Service（FaaS：函数即服务。开发者可以直接将服务业务逻辑代码部署、运行在第三方提供的无状态计算容器中，开发者只需要编写业务代码即可，无须关注服务器，并且代码的执行是由事件触发的。其中 AWS Lambda 是目前最佳的 FaaS 实现之一）。Serverless 是由开发者实现的服务端逻辑，运行在无状态的计算容器中，它是由事件触发，完全被第三方管理的。微信小程序云开发所提供的云函数并不是完整的环境，而是以一个特定的事件为单位的。严格来说，它所提供的云函数功能其实是 FaaS。无服务 Serverless 开发，让开发者不再考虑后端基础设施建设及维护。无服务开发是未来趋势。

开发者可以使用云开发技术开发微信小程序、小游戏，无须搭建服务器，即可使用云端能力。云开发为开发者提供完整的原生云端支持和微信服务支持，弱化后端和运维概念，无须搭建服务器，使用平台提供的 API 进行核心业务开发，即可实现快速上线和迭代，同时这一能力同开发者已经使用的云服务相互兼容，并不互斥。

最初，微信小程序云开发是与微信小程序一起发布的，只能在微信小程序里使用；而现在微信小程序云开发还支持在各个平台（如 Web、H5、微信公众号、iOS、Android 等）中使用，云开发提供的能力和对前端的支持如图 1-1 所示。

图 1-1 云开发提供的能力和对前端的支持

传统开发模式与云开发模式的对比如图 1-2 所示。

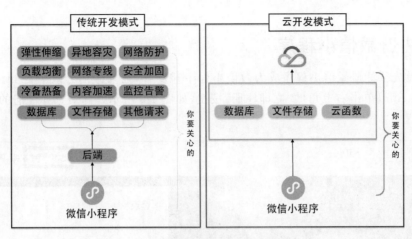

图 1-2 传统开发模式与云开发模式的对比

传统开发模式需要开发者关注后端服务器,比如数据库、文件存储、负载均衡、异地容灾、网络防护等;而云开发模式弱化了后端和运维的概念,只需要关注业务逻辑,就可以完成微信小程序的开发,这样不但大大降低了微信小程序的开发门槛,而且节省了开发者开发部署以及维护的成本。

微信小程序云开发具有如下特点:

(1)免运维:无须自建服务器,免域名备案,支持弹性伸缩;

(2)提升开发和交付效率:一套云开发代码支持 SaaS(Software as a Service)和定制化交付;

(3)简单易用:免鉴权调用微信开放能力和腾讯云高级能力;

(4)与 Web 管理平台无缝对接:支持微信小程序以外的应用访问云开发资源。

1.1.3 微信小程序云开发的核心能力

微信小程序云开发提供了用于支撑完整的后端开发的 4 种基础能力:云数据库、云存储、云函数和云调用,如表 1-1 所示。

表 1-1 云开发提供的基础能力支持

能 力	作 用	说 明
云数据库	无须自建数据库	一个既可在微信小程序前端操作,也能在云函数中读写的 JSON 数据库
云存储	无须自建存储和 CDN(内容分发网络)	在微信小程序前端直接上传/下载云端文件,在云开发控制台可视化管理
云函数	无须自建服务器	在云端运行的代码,微信私有协议天然鉴权,开发者只需编写自身业务逻辑代码
云调用	原生微信服务集成	基于云函数免鉴权使用微信小程序开放接口的能力,包括服务端调用、获取开放数据等能力

1.2 注册微信小程序

用户首先要注册微信小程序成为微信小程序开发者,登录微信公众平台 https://mp.weixin.qq.com,单击右上角的"立即注册"按钮,如图 1-3 所示。随后选择注册的账号类型为微信小程序。

图 1-3 微信公众平台

微信小程序用户注册界面如图 1-4 所示。填写需要注册的邮箱、密码、确认密码和验证码后,单击"注册"按钮。

图 1-4 微信小程序用户注册界面

注意，每个邮箱只能注册一个公众号或微信小程序，所以这一步需要的邮箱必须是之前没有注册过公众号、微信小程序的，而且已经绑定了个人微信号的邮箱也不行。

填写信息并提交注册后，系统会自动发送激活邮件到注册邮箱进行确认，如图 1-5 所示。用户需要登录注册邮箱，单击激活超链接，完成激活操作。

图 1-5 邮箱激活

接下来是完成用户信息登记操作，如图 1-6 所示。这里需要填写主题类型和选择验证方式。主题类型分为个人、企业、政府、媒体和其他组织 5 类，具体信息登记在此不作详述。

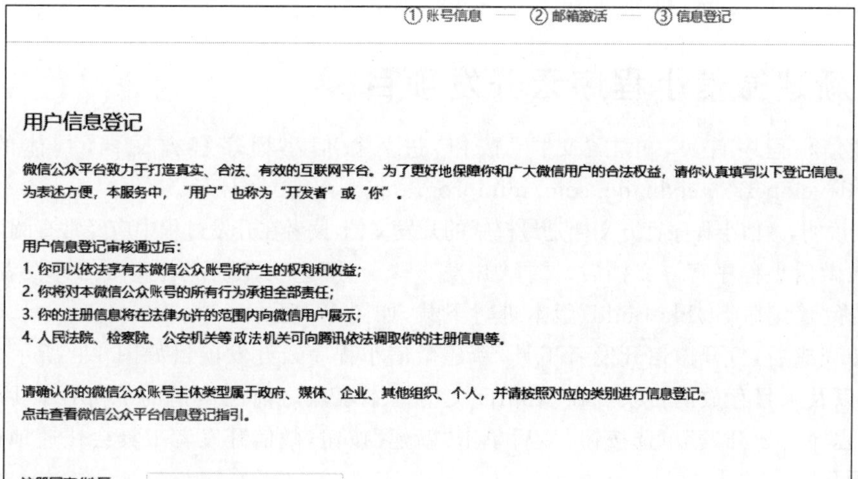

图 1-6 用户信息登记

注意，目前主题类型确认后不支持更改，所以填写时要慎重。

用户根据自己的需求选择主体类型完成注册后，会跳转到微信小程序首页。进入首页后，选择"开发"→"开发设置"选项，可以看到 AppID（微信小程序 ID），这里的 AppID 是新建微信小程序云开发项目时需要用的，开发者需要把这里的 AppID 保存下来，如图 1-7 所示。

图 1-7　微信小程序首页之获取 AppID（微信小程序 ID）

1.3　新建微信小程序云开发项目

在微信小程序首页，单击"文档"按钮，进入微信小程序官方文档，具体的网址为 https://developers.weixin.qq.com/miniprogram/dev/framework。

特别说明，微信小程序官方文档是最权威的开发文档，读者在开发过程中应经常查阅该文档。

进入微信小程序官方文档后，选择"开发"→"工具"选项，随后单击"微信开发者工具"超链接，选择"稳定版 Stable Build"版本进行下载，如图 1-8 所示。

安装完成后，打开微信开发者工具，新建微信小程序云开发项目如图 1-9 所示。填写项目名称，以及项目存放的目录。在 AppID 文本框中填写之前已申请的 AppID，"后端服务"选择"小程序·云开发"单选按钮，最后单击"新建"按钮，微信开发者工具会快速地创建一个云开发项目。

因为项目还没开通云开发环境，此时编译项目，在 console 输出窗口中可以看到错误提示："cloud init error： Error：invalid scope 没有权限,请先开通云服务。"

在使用云开发之前，需要先开通云开发。在微信小程序开发界面，如图 1-10 所示，在工具栏中单击"云开发"按钮。进入云开发控制台界面，如图 1-11 所示，单击"开通"按钮。

图 1-8　微信小程序开发工具下载

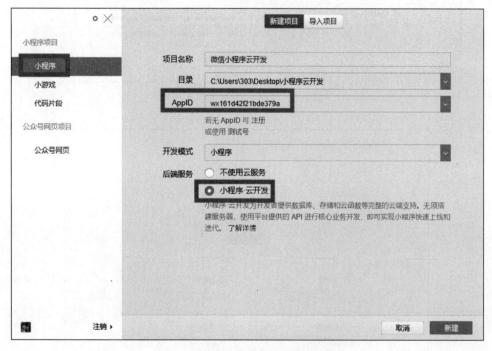

图 1-9　新建微信小程序云开发项目

接下来创建云开发环境，如图 1-12 所示，每个微信小程序账号可以创建两个环境，一般情况下，一个环境用于测试，另一个用于正式上线运行。填写环境名称及环境 ID(可选择默认的)，单击"确定"按钮，即可创建环境。成功创建环境后进入云开发控制台，如图 1-13 所示，可以看到云开发包括(云)数据库、(云)存储和云函数。

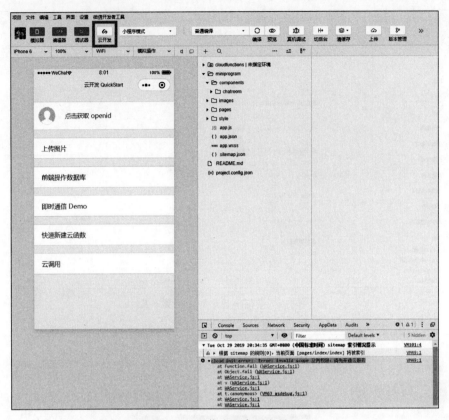

图 1-10 微信小程序开发界面

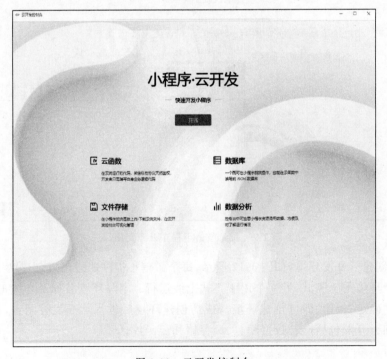

图 1-11 云开发控制台

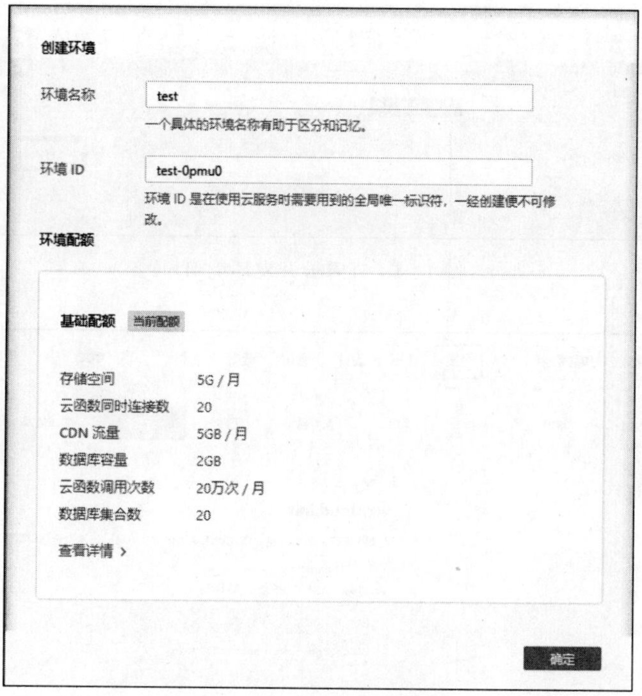

图 1-12 创建云开发环境

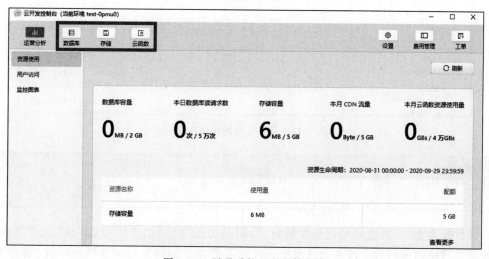

图 1-13 开通后的云开发控制台

在云开发控制台中可以查看云开发环境ID,如图1-14所示,单击右上角的"设置"按钮,在"环境设置"页面中可以看到创建的环境名称和环境ID,这里的环境ID需要在代码中进行配置,记录环境ID的名称。

读者可以进入微信小程序官方文档,查看云开发初始化说明文档,如图1-15所示。选择"开发"→"云开发"选项,在左侧树形目录中选择"SDK 文档"→"初始化"→"小程序",按照文档说明配置wx.cloud.init方法(wx.cloud.init方法在app.js文件中):

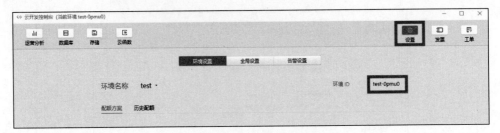

图 1-14　查看云开发环境 ID

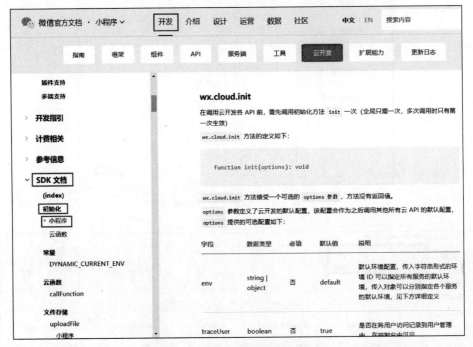

图 1-15　云开发初始化说明文档

```
1  wx.cloud.init({
2    env: 'test-0pmu0'              //环境 ID 获取见图 1-14
3  })
```

到目前为止,云开发环境已经配置好了,但是要正常运行"云开发 QuickStart"项目,还需要把项目中的云函数上传到云开发中。首先需要为云函数根目录选择云环境,如图 1-16 所示,右击 cloudfunctions 目录,选择当前云开发环境,这里选择 test 环境。然后开始上传云函数,如图 1-17 所示,单击云函数,比如在 login 云函数上右击,在弹出的快捷菜单中选择"创建并部署:云端安装依赖(不上传 node_modules)"选项,按照此方法,把目录中的 4 个云函数依次上传云端。

上传完云函数后,在云开发控制台中选择"云函数",就可以看到已经上传的云函数列表,如图 1-18 所示。这样,"云开发 QuickStart"项目就可以正常运行了,读者可以在开发工具中编译该项目,在该项目上体验云数据库、云存储和云函数,这 3 大功能本书将在后续章节中进行详细讲解。

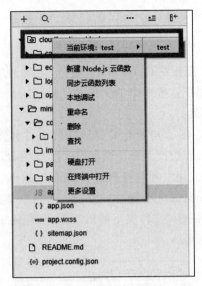

图 1-16　为云函数根目录选择云环境

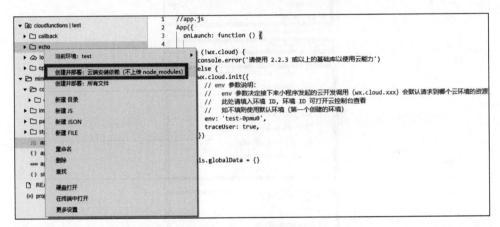

图 1-17　上传云函数

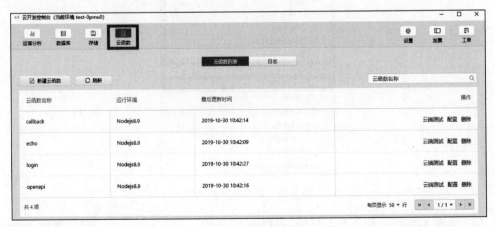

图 1-18　云函数列表

1.4 初始化项目

1.3 节已经生成了一个"云开发 QuickStart"项目,实际上开发微信小程序时,有些文件和代码是不需要的,因此应先把这些多余的文件和代码删除。

项目初始化需要两个步骤:

(1) 删除不需要的文件。项目初始化需要删除的文件如图 1-19 所示。用户可以在需要删除的文件上右击,在弹出的快捷菜单中选择"删除"选项,当然开发者也可以直接在项目文件夹下直接进行删除操作。开发者需要删除 cloudfunctions|test 云函数目录下的所有云函数(callback、echo、login、openapi),删除 miniprogram 下的 components 文件夹(整个 components 文件夹),删除 images 文件夹下的所有图片,删除 pages 文件夹下的所有页面,删除 style 下的样式(guide.wxss)。

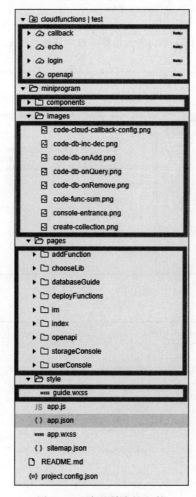

图 1-19 需要删除的文件

(2) 删除 app.json 中多余的代码,如图 1-20 所示。进入 app.json 文件,删除 pages 下所有的页面配置代码,然后增加一条"pages/home/home"。需要注意的是,"pages/index/

index"后面没有逗号。操作完后 pages 代码如下：

```
"pages": [
  "pages/home/home"
],
```

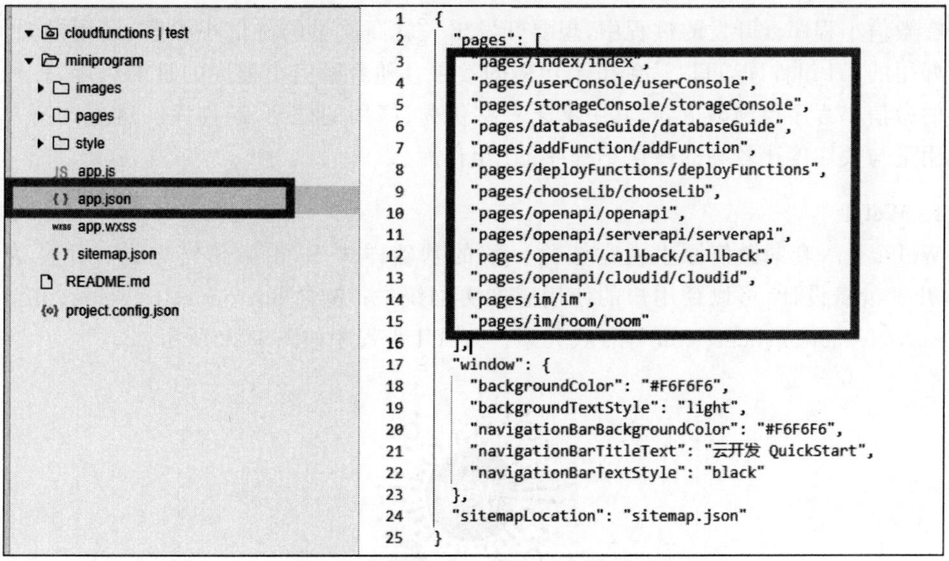

图 1-20　删除 app.json 中多余的代码

单击图 1-21 中的"编译"按钮（或者在任何页面中按 Ctrl＋S 组合键），项目会自动在 pages 下建立 home 页面。至此，项目的初始化就完成了，下面开始编写项目。

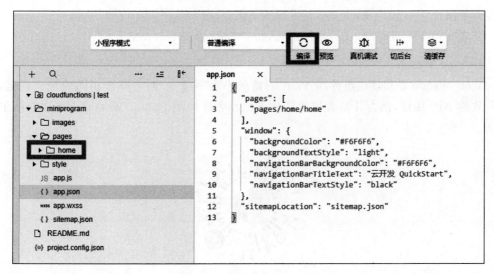

图 1-21　编译生成 home 页面

1.5 微信小程序 UI 组件库及图标库

1.5.1 常用微信小程序 UI 组件库

在微信小程序云开发的过程中，想要开发出一款高质量的微信小程序，运用框架、选择一款好用的 UI 组件库，可以达到事半功倍的效果。随着微信小程序的日渐火爆，各种不同类型的微信小程序也渐渐更新，其中不乏一些优秀、好用的框架/组件库。这里介绍几款用户使用量与关注度比较高的微信小程序 UI 组件库。

1. WeUI

WeUI 是一套同微信原生视觉体验一致的基础样式库，由微信官方设计团队为微信 Web 开发量身设计，可以令用户的使用感知更加统一；包含 button、cell、dialog、progress、toast、article、actionsheet、icon 等各式元素。WeUI 小程序如图 1-22 所示。

图 1-22　WeUI 小程序

GitHub 网址：https://github.com/Tencent/weui。

效果网址：https://weui.io。

开发文档参考网址：https://github.com/Tencent/weui/wiki。

2. Vant Weapp

Vant Weapp 是移动端组件库 Vant 的微信小程序版本，两者基于相同的视觉规范，提供一致的 API 接口，助力开发者快速搭建微信小程序应用。Vant Weapp 小程序如图 1-23 所示。

图 1-23　Vant Weapp 小程序

GitHub 网址：https://github.com/youzan/vant-weapp。

开发文档参考网址：https://youzan.github.io/vant-weapp/#/intro。

3. iView Weapp

iView Weapp 是由 TalkingData 发布的组件库,是一套高质量的微信小程序 UI 组件库。iView Weapp 小程序如图 1-24 所示。

图 1-24　iView Weapp 小程序

GitHub 网址:https://github.com/TalkingData/iview-weapp。

开发文档参考网址:https://weapp.iviewui.com/docs/guide/start。

4. ColorUI

ColorUI 有鲜亮的高饱和色彩,是专注视觉的微信小程序组件库。ColorUI 小程序如图 1-25 所示。

图 1-25　ColorUI 小程序

GitHub 网址:https://github.com/weilanwl/ColorUI。

开发文档参考网址(编辑中):https://www.color-ui.com。

5. Wux Weapp

Wux Weapp 的 UI 样式可配置,拓展灵活,轻松适应不同的设计风格,60 多个丰富的组件,能够满足移动端开发的基本需求。Wux Weapp 小程序如图 1-26 所示。

图 1-26　Wux Weapp 小程序

GitHub 网址:https://github.com/wux-weapp/wux-weapp。

开发文档参考网址:https://wux-weapp.github.io/wux-weapp-docs/#/introduce。

6. TaroUI

TaroUI 是由京东的凹凸实验室倾力打造的多端开发解决方案。现如今市面上端的形态多种多样，Web、ReactNative、微信小程序等各种端大行其道，当业务要求同时在不同的端都要有所表现时，针对不同的端去编写多套代码的成本显然非常高，这时只编写一套代码就能够适配到多端的能力显得极为需要。

使用 TaroUI，可以只书写一套代码，再通过 Taro 的编译工具，将源代码分别编译出可以在不同端（微信小程序、H5、RN 等）运行的代码。TaroUI 小程序如图 1-27 所示。

图 1-27 TaroUI 小程序

GitHub 网址：https://github.com/NervJS/taro-ui。

开发文档参考网址：https://taro-ui.aotu.io/#/docs/introduction。

7. MinUI

MinUI 是基于微信小程序自定义组件特性开发而成的一套简洁、易用、高效的组件库，适用场景广，覆盖微信小程序原生框架、各种微信小程序组件主流框架等。MinUI 小程序如图 1-28 所示。

图 1-28 MinUI 小程序

GitHub 网址：https://github.com/meili/min-cli。

开发文档参考网址：https://meili.github.io/min/docs/minui/index.html#README。

本书重点介绍 ColorUI 组件库和 Vant Weapp 组件库。

1.5.2 ColorUI 组件库

ColorUI 是一个 CSS（层叠样式表）类的 UI 组件库，其组件在美观性方面比较突出，但是目前还没有开发文档，开发人员在官方网站（https://www.color-ui.com/）下载 ColorUI 组件库后解压缩，用微信小程序打开文件中的 demo 项目，就可以浏览、编辑 ColorUI 组件库了。建议在开发阶段开发者用微信小程序打开 ColorUI 组件库，直接从该组件库引用自己需要的样式。

在自己的微信小程序中使用 ColorUI 需要两个步骤：

步骤一：下载源码解压获得 ColorUI-master，复制目录下的 demo\colorui 文件夹到你的微信小程序 miniprogram 文件夹下；

步骤二：在 App.wxss 文件中引入关键 ColorUI 中相应的 CSS 文件。

```
1  @import "colorui/main.wxss";
2  @import "colorui/icon.wxss";
```

配置完 ColorUI 就可以在你的微信小程序中使用 ColorUI 组件库了，具体的 ColorUI 组件库的使用将在后续章节进行介绍，这里介绍如何使用 ColorUI 自定义导航栏。导航栏作为常用组件已做过简单封装，当然也可以直接复制代码结构自己修改，达到个性化的目的。

app.js 获得系统参数信息的代码如下：

```
1  onLaunch: function() {
2      wx.getSystemInfo({
3          success: e =>{
4              this.globalData.StatusBar = e.statusBarHeight;
5              let custom = wx.getMenuButtonBoundingClientRect();
6              this.globalData.Custom = custom;
7              this.globalData.CustomBar = custom.bottom + custom.top - e.statusBarHeight;
8          }
9      })
10 }
```

app.js 获得系统参数信息代码植入位置如图 1-29 所示。wx.getSystemInfo 获取系统信息，statusBarHeight 表示状态栏的高度，单位为 px；wx.getMenuButtonBoundingClientRect 获取菜单按钮（右上角的胶囊状按钮）的布局位置信息。坐标信息以屏幕左上角为原点。

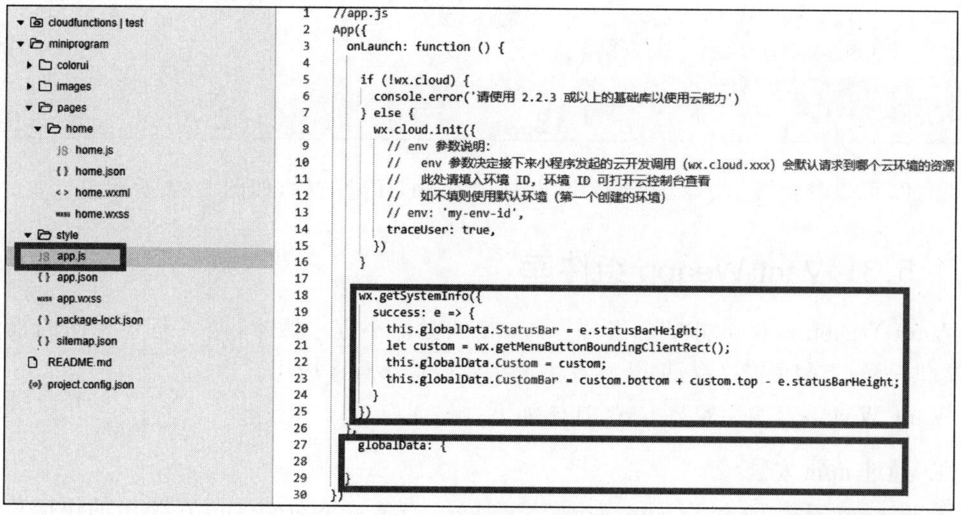

图 1-29 app.js 获得系统参数信息

app.json 配置取消系统导航栏,并全局引入组件,代码如下:

```
1  "window": {
2      "navigationStyle": "custom"
3  },
4  "usingComponents": {
5      "cu-custom":"/colorui/components/cu-custom"
6  }
```

app.json 配置取消系统导航栏代码植入位置如图 1-30 所示。当页面配置 navigationStyle 设置为 custom 时,可以使用第三方组件替代原生导航栏。

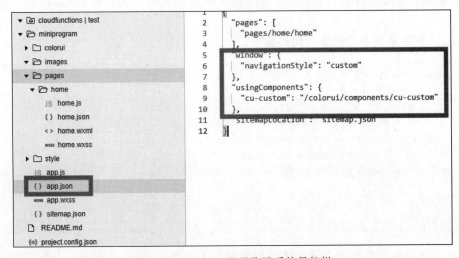

图 1-30　app.json 配置取消系统导航栏

在需要使用导航栏的页面调用 home.wxml,代码如下:

```
1  <cu-custom bgColor="bg-gradual-pink" isBack="{{true}}">
2      <view slot="backText">返回</view>
3      <view slot="content">导航栏</view>
4  </cu-custom>
```

至此,可以看到微信小程序项目中出现了 ColorUI 导航栏,如图 1-31 所示。

1.5.3　Vant Weapp 组件库

Vant Weapp 是移动端 Vue 组件库 Vant 的微信小程序版本,两者基于相同的视觉规范,提供一致的 API 接口,帮助开发者快速搭建微信小程序应用。

Vant Weapp 有两种安装方式,具体如下。

1. 通过 npm 安装

通过 npm 安装 Vant Weapp,如图 1-32 所示,右击 miniprogram,在弹出的快捷菜单中选择"在终端中打开"选项,在打开的终端中通过 npm 方式安装,执行命令:

```
1  npm init    //不运行这条语句,在开发者工具中选择"工具"→"构建 npm"选项会出现:没有找到可
               以构建的 npm 包
2  npm i vant-weapp-S --production
```

图 1-31 ColorUI 导航栏

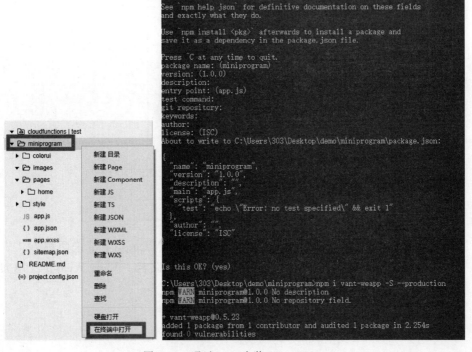

图 1-32 通过 npm 安装 Vant Weapp

接下来开始构建 npm，打开微信开发者工具，选择"工具"→"构建 npm"选项，并勾选"使用 npm 模块"复选框，构建完成后，即可引入组件，如图 1-33 所示。安装完以后在 miniprogram 目录下会生成 miniprogram_npm/vant-weapp 目录。

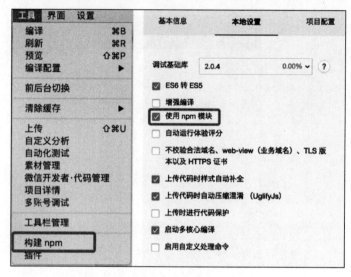

图 1-33　构建 npm

2. 下载代码

直接下载 Vant Weapp 源代码，并将 dist 目录复制到自己项目的 miniprogram 目录下，如图 1-34 所示。Vant Weapp 源代码网址为 https://github.com/youzan/vant-weapp。

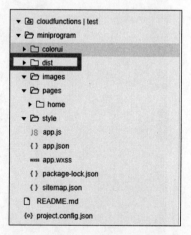

图 1-34　Vant Weapp 复制的目录

读者可以自由选择上面的两种方式之一安装 Vant Weapp，不过这两种方式安装的 vant-weapp 路径略有不同，这里介绍采用第二种方式的路径。Vant Weapp 使用文档见 https://youzan.github.io/vant-weapp/#/quickstart。

下面介绍如何使用 Vant Weapp 按钮组件。

（1）在 home.json 中引入组件，代码如下：

```
1  {
2    "usingComponents": {
3      "van-button": "../../dist/button"
4    }
5  }
```

（2）引入组件后，就可以在 home.wxml 中使用 Vant Weapp 按钮组件了。

```
1   <cu-custom bgColor = "bg-gradual-pink" isBack = "{{true}}">
2       <view slot = "backText">返回</view>
3       <view slot = "content">导航栏</view>
4   </cu-custom>
5
6   <van-button type = "default">默认按钮</van-button>
7   <van-button type = "primary">主要按钮</van-button>
8   <van-button type = "info">信息按钮</van-button>
9   <van-button type = "warning">警告按钮</van-button>
10  <van-button type = "danger">危险按钮</van-button>
```

使用 Vant Weapp 按钮组件显示效果如图 1-35 所示，读者可查阅 Vant Weapp 使用文档，选取自己所需要的组件。

图 1-35　使用 Vant Weapp 按钮组件

1.5.4　iconfont 图标库

在微信小程序开发过程中，会用到各种各样的小图标，而前端开发人员往往不擅长设计这些图标，做出来的项目往往外观不协调。这里介绍 iconfont 图标库，iconfont 是阿里妈妈 MUX 倾力打造的矢量图标管理、交流平台，设计师将图标上传到 iconfont 平台，用户可以自定义下载多种格式的图标，平台也可将图标转换为字体，便于前端工程师自由调整与调用。在没有特殊说明的情况下，本项目使用的小图标均来自 iconfont 图标库 https://www.iconfont.cn。

1.6 微信小程序优化建议

微信小程序分包大小有以下限制：
- 整个微信小程序所有分包大小不超过 16MB；
- 单个分包/主包大小不能超过 2MB。

某些情况下，开发者需要将微信小程序划分成不同的子包，在构建时打包成不同的分包，用户在使用时按需加载，关于分包加载请参考 https://developers.weixin.qq.com/miniprogram/dev/framework/subpackages.html。

在开发微信小程序中的优化建议：

1. 代码包大小的优化

微信小程序开始时代码包限制为 1MB，之后放开了这个限制，增加到 2MB。代码包上限的增加对于开发者来说，能够实现更丰富的功能，但对于用户来说，也增加了下载流量和本地空间的占用。

开发者在实现业务逻辑的同时也有必要尽量减少代码包的大小，因为代码包大小直接影响下载速度，从而影响用户的首次打开体验。

2. 控制代码包内图片资源

微信小程序代码包经过编译后，会放在微信的 CDN 上供用户下载，CDN 开启了 GZIP 压缩，所以用户下载的是压缩后的 GZIP 包，其大小比代码包原体积小。但分析数据发现，不同微信小程序之间的代码包压缩比差异很大，部分可以达到 30%，而部分只有 80%，而造成这部分差异的一个原因就是图片资源的使用。GZIP 对基于文本资源的压缩效果最好，在压缩较大文件时往往可高达 70%～80% 的压缩率，而如果对已经压缩的资源（例如大多数的图片格式）则效果甚微。

3. 及时清理没有使用到的代码和资源

在日常开发的时候，可能会引入一些新的库文件，而过了一段时间后，由于各种原因又不再使用这个库了，这时开发人员常常会只是去掉代码里的引用，而忘记删掉这类库文件。微信小程序打包时会将工程下所有文件都打入代码包内，也就是说，这些没有被实际使用到的库文件和资源也会被打入到代码包里，从而影响整体代码包的大小。

对比上面的 3 条优化建议，这里有几点关于本书的说明。

（1）本书在项目案例中使用 ColorUI 组件库和 Vant Weapp 组件库，但是这两个组件库占用的空间有较大的差别，ColorUI 组件库大小 137KB，Vant Weapp 组件库占用空间 309KB，因此本书优先推荐使用 ColorUI 组件库，如果读者想使用 Vant Weapp 组件库，那么在开发结束后，可以把 Vant Weapp 组件库没有使用的组件删除，以减少代码包的大小。

（2）除了 tabBar 中的图片，其他图片建议放在云存储中，这样不但有利于后期的维护，同时会大大降低代码包的大小，因为图片通常占用较大的空间，但是本书在编写过程中有时为了便于大家自己动手实现书中的效果，会把图片放在本地目录中。

（3）在编写微信小程序的过程中，应该经常关注微信小程序包的大小。把微信小程序包控制在合理的大小范围内，有利于提高微信小程序的用户体验。微信小程序包的大小可以在微信开发者工具中查看，单击"详情"按钮，然后选择"基本信息"子页面，如图 1-36 所示。

图 1-36　查看微信小程序包的大小

第 2 章

微信小程序相关基础知识

2.1 ECMAScript 6 基础知识

2.1.1 ECMAScript 6 简介

微信小程序端采用的是 JavaScript（简称 JS）语言，开发者使用 JavaScript 来开发业务逻辑以及调用微信小程序的 API 来完成业务需求，而云开发使用的云函数现在唯一支持的语言为 Node.js。Node.js 是一个基于 Chrome V8 引擎的 JavaScript 运行环境，使用了高效、轻量级的事件驱动以及非阻塞的 I/O 模型，通常会将 Node.js 作为后端的语言来使用。这里先介绍 JavaScript 和 Node.js 的区别。微信小程序中的 JavaScript 实现是由 ECMAScript、微信小程序架构和微信小程序封装的 API 组成，如图 2-1 所示。它和 Node.js 中的 JavaScript 相比，缺少 Node.js 中的 Native 模块和 NPM，也没办法使用原生库和对 NPM 包管理，与浏览器 JavaScript（ECMAScript/DOM/BOM）相比，没有 DOM（文档对象模型）和 BOM（浏览器对象模型）对象，所以类似于 jQuery 这种浏览器类库是无法使用的。Node.js 中的 JavaScript 实现由 ECMAScript、Native 模块和 NPM 组成，如图 2-2 所示。Native 就是原生的模块，通过这个模块使用 JavaScript 语言本身不具有的一些能力；NPM 是包管理系统，也是目前最大的开原库生态系统。通过各种 NPM 扩展包可以快速地实现

图 2-1 微信小程序中的 JavaScript 实现

一些功能。微信小程序中的 JavaScript 实现和 Node.js 中的 JavaScript 实现的相同点为它们都是基于 ECMAScript 的一种实现。

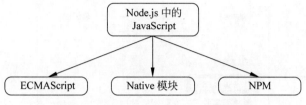

图 2-2　Node.js 中的 JavaScript 实现

ECMAScript 是一种由 Ecma 国际(前身为欧洲计算机制造商协会，European Computer Manufacturers Association)通过 ECMA-262 标准化的脚本程序设计语言。这种语言在万维网上应用广泛，它往往被称为 JavaScript 或 JScript，所以它可以理解为是 JavaScript 的一个标准。ECMAScript 6 简称 ES6，是 JavaScript 语言的下一代标准，是在 2015 年发布的，所以 ES6 也称为 ECMAScript 2015。ECMAScript 和 JavaScript 的关系是，前者是后者的规格，后者是前者的一种实现(另外的 ECMAScript 语言还有 Jscript 和 ActionScript)。在日常场合，这两个词是可以互换的。

微信小程序可以运行在 3 大平台上：iOS 平台，包括 iOS9、iOS10、iOS11、iOS13；Android 平台；微信小程序 IDE。其区别主要是体现 3 大平台实现的 ECMAScript 的标准有所不同。截至目前一共有 7 个版本的 ECMAScript 标准，开发者大部分使用的是 ES5 和 ES6 的标准，但是在微信小程序中，iOS9 和 iOS10 所使用的运行环境并没有完全兼容到 ES6 标准，一些 ES6 中规定的语法和关键字是没有的或者与标准有所不同，所以一些开发者会发现有些代码在旧的手机操作系统上出现一些语法错误。为了帮助开发者解决这类问题，微信小程序 IDE 提供语法转码工具，帮助开发者将 ES6 代码转为 ES5 代码，从而在所有的环境都能得到很好的执行。开发者需要单击微信开发者工具右上角的"详情"按钮，选择"本地设置"子页面，勾选"ES6 转 ES5"复选框，开启此功能，如图 2-3 所示。

本章介绍 ES6 的语法，因为微信小程序端和云函数都是采用 ES6 语法，所以 ES6 的语法在微信小程序端和云函数上均可使用。采用 VS Code 编译器，首先需要安装 Node.js(当然也可以直接使用微信小程序开发者工具)。进入 Node.js 官方网站(https://nodejs.org/en/download/)，以 Windows 系统为例，下载 Windows 安装包(.msi)，如图 2-4 所示，下载后直接安装该软件。

Node.js 软件安装比较简单，默认安装就可以了，安装后进入 cmd 控制台，输入 node -v 可以查看 Node.js 的版本信息，表示 Node.js 安装成功，最新版的 Node.js 在安装的同时也安装了 npm，执行 npm -v 查看 npm 版本，如图 2-5 所示。

本章使用 VS Code 演示 ES6 的语法，读者需要自行安装 VS Code，安装完 VS Code 后，新建项目并新建.js 文件，在文件中输入：

```
1  var name = 'Hello world!'
2  console.log(name)
```

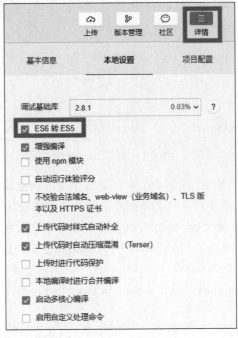

图 2-3　勾选"ES6 转 ES5"复选框

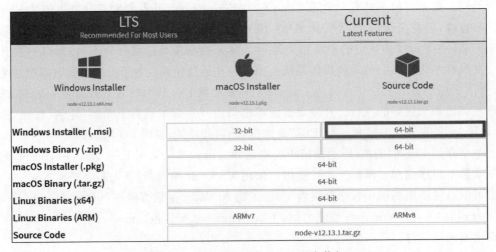

图 2-4　官方网站下载 Node.js 安装包

图 2-5　查看版本信息

在 VS Code 调试菜单中选择"启动调试"选项,随后选择 Node.js 环境,点击"创建 launch.json 文件",如图 2-6 所示,再次选择 Node.js 环境,这时会在调试控制台输出"Hello world!",表示 VS Code 运行 Node.js 环境成功。

图 2-6 创建 launch.json 文件

本章侧重于介绍微信小程序和云函数中常用的 ES6 语法,至于 npm 的使用会在后续的章节介绍。

2.1.2 ECMAScript 变量

ECMAScript 中的变量无特定的类型,定义变量时只用 var 运算符,可以将它初始化为任意值,例如:

```
1  var color = "red";
2  var num = 25;
3  var visible = true;
4  console.log("color:", color)
5  console.log("num:", num)
6  console.log("visible:", visible)
```

输出结果为:

```
color: red
num: 25
visible: true
```

每行结尾的分号可有可无,ECMAScript 允许开发者自行决定是否以分号结束一行代码。如果没有分号,ECMAScript 就把折行代码的结尾看作该语句的结尾。

在微信小程序中定义变量除了使用 var 以外,还经常看到 let 和 const,ES6 新增了 let 和 const 命令,用来声明变量。let 的用法类似于 var,但是所声明的变量只在 let 命令所在的代码块内有效。例如:

```
1  {
2      let a = 10;
3      var b = 1;
4  }
```

```
5    a//ReferenceError: a is not defined
6    b//1
```

上面代码在代码块中分别用 let 和 var 声明了两个变量。然后在代码块外调用这两个变量,结果 let 声明的变量报错,var 声明的变量返回了正确的值。这表明,let 声明的变量只在它所在的代码块有效。

只要块级作用域内存在 let 命令,它所声明的变量就"绑定"(binding)这个区域,不再受外部的影响。

```
1    var tmp = 123;
2    if (true) {
3        tmp = 'abc'; // ReferenceError: Cannot access 'tmp' before initialization
4        let tmp;
5    }
```

上面代码中,存在全局变量 tmp,但是块级作用域内 let 又声明了一个局部变量 tmp,导致后者绑定这个块级作用域,所以,在 let 声明变量前对 tmp 赋值会报错。ES6 明确规定,如果区块中存在 let 和 const 命令,这个区块对这些命令声明的变量,从一开始就形成了封闭作用域。凡是在声明之前就使用这些变量就会报错。总之,在代码块内,使用 let 命令声明变量之前,该变量都是不可用的。这在语法上称为"暂时性死区"。暂时性死区的本质就是,只要一进入当前作用域,所要使用的变量就已经存在了,但是不可获取,只有等到声明变量的那一行代码出现,才可以获取和使用该变量。

const 声明一个只读的常量。一旦声明,常量的值就不能改变。例如:

```
1    const PI = 3.1415;
2    PI        // 3.1415
3    PI = 3; // TypeError: Assignment to constant variable.
```

上面代码表明改变常量的值会报错。const 声明的变量不得改变值。这意味着 const 一旦声明变量,就必须立即初始化,不能留到以后赋值。比如定义常量 PI,会出现错误 SyntaxError: Missing initializer in const declaration。对于 const 来说,只声明不赋值,就会报错。此外 const 的作用域与 let 命令相同:只在声明所在的块级作用域内有效;const 命令同样存在暂时性死区,只能在声明的位置后面使用。

在微信小程序开发过程中,通常需要查看一些变量的值,虽然微信开发者工具提供了 AppData 窗口可以看到每个页面中 data 的值,但是有时需要看一些变量运行中值的变化,而这些变量又不需要在页面上显示(不需要保存到页面的 data 数据中),这时最常用的方法是使用 console.log() 打印变量,如果是在微信小程序中使用 console.log() 输出,就会在 Console 窗口中显示信息,如果是在云函数中使用 console.log(),这时就需要进入云开发控制台,选择"云函数"→"日志"选项进行查看。

一般情况下用来输出信息的方法主要是如下 4 个:

(1) console.log():用于输出普通信息。

(2) console.info():用于输出提示性信息。

(3) console.error()：用于输出错误信息。
(4) console.warn()：用于输出警示信息。
(5) console.debug()：用于输出调试信息。

在微信小程序中演示如下：

```
onLoad: function (options) {
  console.log('aaa')              //用于输出普通信息
  console.info('aaa')             //用于输出提示性信息
  console.error('aaa')            //用于输出错误信息
  console.warn('aaa')             //用于输出警示信息
},
```

在微信开发者工具的 Console 窗口输出的信息如图 2-7 所示。从图 2-7 中可以看出，console.log()和 console.info()没什么区别，console.log()的作用是向控制台输出一条信息，它包括一个指向该行代码位置的超链接（直接点击该超链接，可以跳转到该代码行）。console.info()的作用是向控制台输出一条信息，该信息包含一个表示"信息"的图标（实际上微信开发者工具中在 Console 窗口中并没有显示"信息"图标）和指向该行代码位置的超链接。console.error()的作用是向控制台输出一条信息，该信息包含一个表示"错误"的图标。console.warm()的作用是向控制台输出一条信息，该信息包含一个表示"警告"的图标。在实际使用中 console.log()的用处是最大的，用户在写代码的过程中要充分利用 console.log()的功能来调试程序。

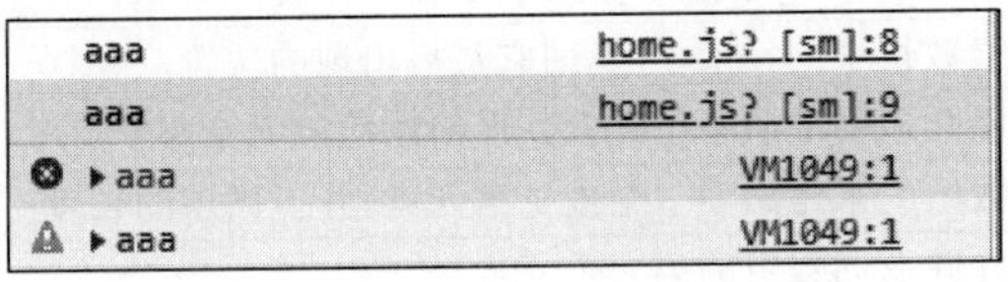

图 2-7　Console 窗口输出信息

console.log()支持 C 语言 printf 式的格式化输出，当然，也可以不使用格式化输出达到同样的目的，例如：

```
var animal = 'frog', count = 10;
console.log("The %s jumped over %d tall buildings", animal, count);
console.log("The", animal, "jumped over", count, "tall buildings");
```

输出结果为：

```
The frog jumped over 10 tall buildings
The frog jumped over 10 tall buildings
```

当然 console.log()也支持微信小程序中最常用的数组和对象，例如：

```
var items = ['塞向秋','瑞访天','乐晏'];
var person = { "name":"塞向秋", "sex":"男", "mobile":"15888837968", "searchfield":
"jianxiangqiujxq 塞向秋" }
console.log(items)
console.log(person)
```

输出结果为:

```
Array(3) ["塞向秋","瑞访天","乐晏"]
Object {name: "塞向秋", sex: "男", mobile: "15888837968", searchfield: "jianxiangqiujxq 塞向秋"}
```

这里需要说明的是,console.log()的用法相对比较简单,但又是编写微信小程序端逻辑代码和云函数最常用的语句,配合使用console.log()可以在程序运行过程中查看变量的状态,便于调试程序。

2.1.3 ECMAScript 数组和对象

数组用于在变量中存储多个值,例如:

```
1  var items = ['塞向秋', '瑞访天', '乐晏'];
2  console.log(items[0], items[1], items[2]);       //塞向秋 瑞访天 乐晏
```

数组中第一个数组元素的索引值为 0,第二个索引值为 1,以此类推。

这里列举了微信小程序开发中数组的常见操作,如表 2-1 所示。

表 2-1 数组的常见操作

命令	说明
concat()	连接两个或更多的数组,并返回结果
indexOf()	搜索数组中的元素,并返回它所在的位置
lastIndexOf()	搜索数组中的元素,并返回它最后出现的位置
pop()	删除数组的最后一个元素并返回删除的元素
push()	向数组的末尾添加一个或更多元素,并返回新的长度
shift()	删除并返回数组的第一个元素
unshift()	向数组的开头添加一个或更多元素,并返回新的长度
splice()	从数组中添加或删除元素
toString()	把数组转换为字符串,并返回结果
length()	设置或返回数组元素的个数
map()	返回一个新数组,数组中的元素为原始数组元素调用函数处理后的值

这里简单演示一下这些操作:

```
1   var items = ['塞向秋', '瑞访天', '乐晏'];
2   console.log(items.length);           //3
3   items.push('果明');
4   console.log(items);                  //Array(4) ["塞向秋","瑞访天","乐晏","果明"]
5   var index = items.indexOf("瑞访天");
6   console.log(items[index]);           //"瑞访天"
7   var item = items.pop()
8   console.log(items);                  //Array(3) ["塞向秋","瑞访天","乐晏"]
9   item = items.unshift('经语心');
10  console.log(items);                  //["经语心","塞向秋","瑞访天","乐晏"]
11  items.splice(2, 1);                  //删除数组下标为 2 的元素
12  console.log(items);                  //["经语心","塞向秋","乐晏"]
```

输出结果为：

```
3
Array(4) ["塞向秋","瑞访天","乐晏","果明"]
瑞访天
Array(3) ["塞向秋","瑞访天","乐晏"]
Array(4) ["经语心","塞向秋","瑞访天","乐晏"]
Array(3) ["经语心","塞向秋","乐晏"]
```

这里需要特别说明一下数组的 push() 和 concat() 操作的区别，push() 遇到数组参数时，把整个数组参数作为一个元素，而 concat() 则是拆开数组参数，将元素逐个加进去；push() 直接改变当前数组，而 concat 不改变当前数组。接下来演示两者之间的区别。

```
1   var arr = [];
2   arr.push(1);
3   arr.push(2);
4   arr.push([3, 4])
5   arr.push(5, 6);
6   arr = arr.concat(7);
7   arr = arr.concat([8, 9]);
8   arr = arr.concat(10, 11);
9   for (var i in arr) {
10      console.log(i + "-----" + arr[i]);
11  }
12  console.log(arr)
```

输出结果为：

```
0-----1
1-----2
2-----3,4
3-----5
4-----6
5-----7
6-----8
7-----9
8-----10
9-----11
Array(10) [1, 2, Array(2), 5, 6, 7, 8, 9, …]
```

在代码第 4 行，push() 遇到数组参数时，把整个数组参数作为一个元素；在代码第 7 行，concat() 则是拆开数组参数，将元素逐个加进去。

接下来演示 map() 方法的用法。map() 方法返回一个新数组，数组中的元素为原始数组元素调用函数处理后的值，同时不会改变原来的数组。比如：

```
1   const arr = [1, 3, 4, 5, 6, 7, 8, 10];
2   const cube = (num) =>{
3       return num * num * num;
```

```
4    }
5    const res = arr.map(cube)
6    console.log(arr)
7    console.log(res)
```

输出结果为：

```
Array(8) [1, 3, 4, 5, 6, 7, 8, 10]
Array(8) [1, 27, 64, 125, 216, 343, 512, 1000]
```

第 5 行中，调用函数 cube 没有传递参数，因为本质上是用元素作为函数参数去调用函数，所以无须加上参数。

对象也是变量，但是对象包含很多值。例如：

```
{"name":"塞向秋","sex":"男","mobile":"15888837968"}
```

JSON(JavaScript Object Notation)是一种轻量级的数据交换格式。微信小程序和后端之间的交互都是采用 JSON 格式的数据。这里所说的对象就是指 JSON 格式的数据。对象必须遵守如下写法：对象被花括号{}包围；对象以键值对书写；键必须是字符串，值必须是有效的 JSON 数据类型(字符串、数字、对象、数组、布尔或 null)；键和值由冒号分隔；每个键值对由逗号分隔。

有两种方式可以访问对象属性。

方式一：通过使用点号(.)来访问对象值。

```
objectName.propertyName
```

方式二：使用方括号[]来访问对象值。

```
objectName["propertyName"]
```

例如：

```
1    var person = { "name":"塞向秋", "sex":"男", "mobile":"15888837968", "searchfield":
2    "jianxiangqiujxq 塞向秋" }
3    console.log(person.name)              //塞向秋
4    console.log(person["name"])           //塞向秋
5    console.log(person)
6    //Object {name:"塞向秋", sex:"男", mobile:"15888837968", searchfield:"jianxiangqiujxq 塞向秋"}
```

用这两种方式访问对象属性一般来说没有什么区别，一般采用方式一 objectName.propertyName 这种写法，如果字段名称是中文，则在微信小程序 Wxml 页面访问对象就必须使用方法二，例如 person["姓名"]。

可以通过使用 for…in 遍历对象属性，也可以在 for…in 循环中使用括号标记法来访问属性值，例如：

```
1   var person = { "name": "蹇向秋", "sex": "男", "mobile": "15888837968", "searchfield":
2   "jianxiangqiujxq 蹇向秋" }
3   for (att in person) {
4       console.log(att, ":", person[att])
5   }
6   person.sex = "女"
7   delete person.searchfield
8   console.log(person)
```

输出结果为：

```
name : 蹇向秋
sex : 男
mobile : 15888837968
searchfield : jianxiangqiujxq 蹇向秋
Object {name: "蹇向秋", sex: "女", mobile: "15888837968"}
```

代码第 3～5 行使用 for…in 遍历对象属性并使用括号标记法来访问属性值；代码第 6 行使用点号来修改对象中的值；代码第 7 行使用 delete 关键词来删除对象的属性。

微信小程序云开发所使用的数据库本质上就是一个 MongoDB 数据库，因此微信小程序和数据库进行交互时都是采用 JSON 格式，当有多个记录时，数据库返回的数据通常是对象数组，也就是说数组的元素可以是对象。例如：

```
[{"name":"蹇向秋","sex":"男","mobile":"15888837968"},
 {"name":"犹淑慧","sex":"女","mobile":"13506897988"},
 {"name":"睦琪睿","sex":"女","mobile":"18762735852"},
 {"name":"瑞访天","sex":"男","mobile":"18257507459"}]
```

这时可以结合数组和对象的方法进行访问，例如：

```
1   var records = [
2       { "name": "蹇向秋", "sex": "男", "mobile": "15888837968"},
3       { "name": "犹淑慧", "sex": "女", "mobile": "13506897988"},
4       { "name": "睦琪睿", "sex": "女", "mobile": "18762735852"},
5       { "name": "瑞访天", "sex": "男", "mobile": "18257507459"}
6   ]
7   console.log(records[0])
8   console.log(records[0].name)
9   console.log(records[0]["mobile"])
```

输出结果为：

```
Object {name: "蹇向秋", sex: "男", mobile: "15888837968"}
蹇向秋
15888837968
```

在数据传输流程中，JSON 格式的数据是以文本即字符串的形式传递的，而微信小程序操作的是 JSON 对象，所以 JSON 对象和 JSON 字符串之间的相互转换也是很常见的操作。

1. JSON 对象转换为字符串

使用 JSON.stringify() 把 JSON 对象转换为字符串，例如：

```javascript
var person_json = { "name": "塞向秋", "sex": "男", "mobile": "15888837968" }
var person_str = JSON.stringify(person_json)
console.log(person_json)
console.log(person_str)
```

输出结果为：

```
Object {name: "塞向秋", sex: "男", mobile: "15888837968"}
{"name":"塞向秋","sex":"男","mobile":"15888837968"}
```

2. 字符串转换为 JSON 对象

使用 JSON.parse() 把字符串转换为 JSON 对象，例如：

```javascript
var person_str = '{ "name": "塞向秋", "sex": "男", "mobile": "15888837968" }'
var person_json = JSON.parse(person_str)
console.log(person_str)
console.log(person_json)
```

输出结果为：

```
{ "name": "塞向秋", "sex": "男", "mobile": "15888837968" }
Object {name: "塞向秋", sex: "男", mobile: "15888837968"}
```

JSON.stringify() 和 JSON.parse() 通常会出现在页面跳转时传递数组、对象参数。例如 list 是一个对象数组，需要把 list 作为参数传递给下一个页面，页面跳转如下：

```javascript
wx.navigateTo({
  url: '../order/order?orderinfo=' + JSON.stringify(list)
})
```

这里通过 url 传值把 list 数组对象转换为字符串传递给下一个页面。相应地，下一个页面需要把字符串转换为对象，例如：

```javascript
var orderinfo = JSON.parse(options.orderinfo)
```

有时页面传递的字符串参数含有特殊字符（;、/、?、:、@、&、=、+、$、#之类的字符），这时就需要把传递的参数用 encodeURIComponent() 把字符串作为 URI 组件进行编码，在接收参数页面，将参数用 decodeURIComponent() 解析后，再使用 JSON.parse() 将字符串转为对象，例如页面传递为 encodeURIComponent(JSON.stringify(list))，页面接收为 JSON.parse(decodeURIComponent(options.orderinfo))。

2.1.4　ECMAScript 语句

1. if 语句

if 语句是 ECMAScript 中最常用的语句之一,事实上在许多计算机语言中都是如此。if 语句的语法如下:

```
1  if(condition)
2      statement1
3  else
4      statement2
```

其中,condition 可以是任意表达式,计算的结果甚至不必是真正的 boolean 值,ECMAScript 会把它转换为 boolean 值。如果条件计算结果为 true,则执行 statement1;如果条件计算结果为 false,则执行 statement2。每个语句都可以是单行代码,也可以是代码块。

2. 迭代语句

迭代语句可以使用 for 语句和 while 语句,这里以 for 语句为例讲解迭代语句。for 语句是前测试循环,而且在进入循环之前,能够初始化变量,并定义循环后要执行的代码。例如:

```
1  var items = ['寨向秋','瑞访天','乐晏'];
2  for (var i = 0; i< items.length; i++)
3      console.log(items[i])
```

输出结果为:

```
寨向秋
瑞访天
乐晏
```

除了使用 for 语句外,还可以使用 for…in 语句。例如:

```
1   var items = ['寨向秋','瑞访天','乐晏'];
2   for (var index in items){
3       console.log(items[index])
4   }
5
6   var person = { "name":"寨向秋","sex":"男","mobile":"15888837968","searchfield":
7   "jianxiangqiujxq 寨向秋" }
8   for (var att in person){
9       console.log(att,":", person[att])
10  }
```

输出结果为：

```
塞向秋
瑞访天
乐晏
name：塞向秋
sex：男
mobile：15888837968
searchfield：jianxiangqiujxq 塞向秋
```

代码第 1~4 行演示了数组中使用 for…in 语句，代码第 6~10 行演示了对象中使用 for…in 语句，可见 for…in 语句不但能遍历数组，而且能遍历对象。除了使用 for…in 语句外还可以使用 for…of 语句。例如：

```
1  var items = ['塞向秋','瑞访天','乐晏'];
2  for (var value of items) {
3      console.log(value)
4  }
```

输出结果为：

```
塞向秋
瑞访天
乐晏
```

其中，for…in 语句是 ES5 标准，遍历的是 key（可遍历对象、数组或字符串的 key）；for…of 是 ES6 标准，遍历的是 value。for…of 语句循环时不会循环对象的 key，只会循环出数组的 value，因此 for…of 语句不能循环遍历普通对象，对普通对象的属性遍历推荐使用 for…in 语句。如果实在想用 for…of 语句来遍历普通对象的属性，可以通过和 Object.keys() 搭配使用，先获取对象的所有 key 数组然后遍历。例如：

```
1  var person = { "name": "塞向秋", "sex": "男", "mobile": "15888837968", "searchfield":
2  "jianxiangqiujxq 塞向秋" }
3  for (var att of Object.keys(person)) {
4      console.log(person[att])
5  }
```

输出结果为：

```
塞向秋
男
15888837968
jianxiangqiujxq 塞向秋
```

如果是数组还可以使用 forEach() 方法，forEach() 方法为每个数组元素调用一次函数（回调函数）。例如：

```
1  var items = ['塞向秋', '瑞访天', '乐晏'];
2  items.forEach(function (value) {
3      console.log(value)
4  })
```

输出结果为：

```
塞向秋
瑞访天
乐晏
```

2.1.5　ECMAScript 6 异步操作和 Async 函数

　　JavaScript 是单进程执行的，同步操作会对程序的执行进行阻塞处理。所以，异步编程对 JavaScript 语言很重要。Javascript 语言的执行环境是单线程的，如果没有异步编程，根本没法用，非卡死不可。所谓异步，简单说就是一个任务分成两段，先执行第一段，然后转而执行其他任务，等做好了准备，再回过头执行第二段。比如，有一个任务是读取文件进行处理，任务的第一段是向操作系统发出请求，要求读取文件。然后，程序执行其他任务，等到操作系统返回文件，再接着执行任务的第二段（处理文件）。这种不连续的执行，就叫作异步。相应地，连续的执行就叫作同步。由于是连续执行，不能插入其他任务，所以操作系统从硬盘读取文件的这段时间，程序只能干等着。然而开发过程中有时需要把异步操作变成同步操作，这时就需要用到 Promise。

　　Promise 是异步编程的一种解决方案，比传统的解决方案——回调函数和事件——更合理和更强大。它由社区最早提出和实现，ES6 将其写进了语言标准，统一了用法，原生提供了 Promise 对象。所谓 Promise，简单说就是一个容器，里面保存着某个未来才会结束的事件（通常是一个异步操作）的结果。从语法上说，Promise 是一个对象，从它可以获取异步操作的消息。Promise 提供统一的 API，各种异步操作都可以用同样的方法进行处理。

　　Promise 对象有以下两个特点。

　　（1）对象的状态不受外界影响。Promise 对象代表一个异步操作，有 3 种状态：Pending（进行中）、Resolved（已完成，又称 Fulfilled）和 Rejected（已失败）。只有异步操作的结果，可以决定当前是哪一种状态，任何其他操作都无法改变这个状态。这也是 Promise 这个名字的由来，它的英语意思就是"承诺"，表示其他手段无法改变。

　　（2）一旦状态改变，就不会再变，任何时候都可以得到这个结果。Promise 对象的状态改变，只有两种可能：从 Pending 变为 Resolved 和从 Pending 变为 Rejected。只要这两种情况发生，状态就凝固了，不会再变了，会一直保持这个结果。就算改变已经发生了，再对 Promise 对象添加回调函数，也会立即得到这个结果。这与事件（event）完全不同，事件的特点是，如果你错过了它，再去监听，是得不到结果的。

　　有了 Promise 对象，就可以将异步操作以同步操作的流程表达出来，避免了层层嵌套的回调函数。此外，Promise 对象提供统一的接口，使得控制异步操作更加容易。Promise 也有一些缺点。首先，无法取消 Promise，一旦新建它就会立即执行，无法中途取消。其次，如果不设置回调函数，Promise 内部抛出的错误不会反映到外部。最后，当处于 Pending 状态时，无法得知目前进展到哪一个阶段(刚刚开始还是即将完成)。

ES6 规定，Promise 对象是一个构造函数，用来生成 Promise 实例。下面代码创造了一个 Promise 实例：

```
1   let promise = new Promise(function (resolve, reject) {
2       // ...
3       if (/* 异步操作成功 */) {
4           resolve(value);
5       } else {
6           reject(error);
7       }
8   });
9
10  promise.then(res =>{
11      console.log(res)
12  }, err =>{
13      cosole.log(err)
14  })
```

Promise 构造函数接收一个函数作为参数，该函数的两个参数分别是 resolve 和 reject。它们是两个函数，由 JavaScript 引擎提供，不用自己部署。resolve 函数的作用是，将 Promise 对象的状态从"进行中"变为"成功"(即从 Pending 变为 Resolved)，在异步操作成功时调用，并将异步操作的结果作为参数传递出去；reject 函数的作用是，将 Promise 对象的状态从"进行中"变为"失败"(即从 Pending 变为 Rejected)，在异步操作失败时调用，并将异步操作报出的错误作为参数传递出去。Promise 实例生成以后，可以用 then() 方法分别指定 Resolved 状态和 Rejected 状态的回调函数。then() 方法可以接收两个回调函数作为参数：第一个回调函数是在 Promise 对象的状态变为 Resolved 时调用；第二个回调函数是在 Promise 对象的状态变为 Rejected 时调用。其中，第二个函数是可选的，不一定要提供。这两个函数都接收 Promise 对象传出的值作为参数。

Promise 新建后就会立即执行。例如：

```
1   let promise = new Promise(function (resolve, reject) {
2       console.log('Promise');
3       resolve();
4   });
5
6   promise.then(function () {
7       console.log('Resolved.');
8   });
9
10  console.log('Hi!');
```

输出结果为：

```
Promise
Hi!
Resolved.
```

上面代码中,Promise 新建后立即执行,所以首先输出的是 Promise。然后,then()方法指定的回调函数将在当前脚本所有同步任务都执行完后才会执行,所以"Resolved."最后输出。

目前微信小程序针对需要连续执行的任务,提供了回调函数 Callback 风格和 Promise 风格。如果使用 Callback 风格,当需要依次执行多个任务时,就会出现多重嵌套。代码不是纵向发展,而是横向发展,很快就会乱成一团,无法管理。这种情况就称为"回调函数噩梦"(callback hell)。因此在使用微信小程序时建议采用 Promise 风格。

如果希望多个任务能够并发执行,等所有任务执行完进行数据汇总,这时 Promise 可以配合 Promise.all()方法。Promise.all()可以将多个 Promise 实例包装成一个新的 Promise 实例。同时,成功和失败的返回值是不同的,成功时返回的是一个结果数组,而失败时则返回最先被 reject 失败状态的值。例如:

```
1   let p1 = new Promise((resolve, reject) =>{
2       resolve('成功了')
3   })
4
5   let p2 = new Promise((resolve, reject) =>{
6       resolve('success')
7   })
8
9   let p3 = Promise.reject('失败')
10
11  Promise.all([p1, p2]).then((result) =>{
12      console.log(result)            //['成功了', 'success']
13  }).catch((error) =>{
14      console.log(error)
15  })
16
17  Promise.all([p1, p3, p2]).then((result) =>{
18      console.log(result)
19  }).catch((error) =>{
20      console.log(error)             //失败了,输出 '失败'
21  })
```

Promise.all()获得的成功结果的数组里面的数据顺序和 Promise.all()接收到的数组顺序是一致的,即 p1 的结果在前,即便 p1 的结果获取的比 p2 要晚。这带来了一个极大的好处:在前端开发请求数据的过程中,偶尔会遇到发送多个请求并根据请求顺序获取和使用数据的场景,使用 Promise.all()毫无疑问可以解决这个问题。例如:

```
1   let wake = (time) =>{
2       return new Promise((resolve, reject) =>{
3           setTimeout(() =>{
4               resolve(`${time / 1000}秒后醒来`)
5           }, time)
6       })
7   }
```

```
 8
 9   let p1 = wake(3000)
10   let p2 = wake(2000)
11
12   Promise.all([p1, p2]).then((result) =>{
13       console.log(result)              // ['3秒后醒来', '2秒后醒来']
14   }).catch((error) =>{
15       console.log(error)
16   })
```

上述任务 p2 是要比任务 p1 早完成，但是 Promise.all() 获得的成功结果却是先输出 p1 任务的结果，因为 Promise.all() 获得的成功结果的数组里面的数据顺序和 Promise.all() 接收到的数组顺序是一致的。

Promise 的方式虽然解决了"回调函数噩梦"，但是这种方式充满了 Promise 的 then() 方法，如果处理流程复杂，则整段代码将充满 then。语义化不明显，代码流程不能很好地表示执行流程。为了解决 Promise 的问题，ES7 中引入了 async-await 来改造 Promise，async-await 是 ES7 提出的新技术，可以将 Promise 对象中异步的方法以同步的方式书写，减少代码量。其中 async 用来修饰异步代码所在的函数，await 用来修改异步代码。async-await 是 ES7 的语法，到目前为止，微信小程序端默认是不支持 async-await 语法的，需要使用第三方库才能使用，但是云函数已经支持 async-await，因此读者在使用云函数时会看到 async-await。在微信小程序云开发时，微信小程序端建议使用 Promise 方式实现同步操作，在云函数中 async-await 和 Promise 方式都可以使用。云函数目前都支持两种写法，分别是 Promise 和 async-await。async-await 本质上是基于 Promise 的一种语法糖，它只是把 Promise 转换成同步的写法而已。async 出现在方法的声明里面，任何一个方法都可以增加 async。await 放在 task 前面，async 和 await 是成对出现的，只有 async 是没有意义的，只有 await 是会报错的。

async 函数完全可以看作多个异步操作，包装成的一个 Promise 对象，而 await 命令就是内部 then 命令的语法糖。async 函数返回的 Promise 对象，必须等到内部所有 await 命令的 Promise 对象都执行完，才会发生状态改变。也就是说，只有 async 函数内部的异步操作执行完，才会执行 then() 方法指定的回调函数。如果 async 函数中有多个 await 语句，只要一个 await 语句后面的 Promise 变为 reject，那么整个 async 函数都会中断执行。

例如：

```
1   async function f() {
2       await Promise.reject('出错了');
3       await Promise.resolve('hello world'); //不会执行
4   }
5
6   f().then(v => console.log(v))
7    .catch(e => console.log(e))
8   //出错了
```

上面代码中，第二个 await 语句是不会执行的，因为第一个 await 语句状态变成了 reject。为了避免出现这种问题，可以将第一个 await 放在 try…catch 结构里面，这样第二个 await 就会执行。

```
1  async function f() {
2      try {
3          await Promise.reject('出错了');
4      } catch (e) {
5      }
6      return await Promise.resolve('hello world');
7  }
8
9  f().then(v => console.log(v))
10     .catch(e => console.log(e))
11 // hello world
```

await 命令后面的 Promise 对象，运行结果可能是 Rejected，所以最好把 await 命令放在 try…catch 代码块中；如果有多个 await 命令，可以统一放在 try…catch 结构中。

2.2 微信小程序框架

2.2.1 注册页面的使用

对于微信小程序中的每个页面，都需要在页面对应的 JS 文件中进行注册，指定页面的初始数据、生命周期回调、事件处理函数等。使用 Page 构造器注册页面，页面使用 Page() 进行构造。Page 接收一个 Object 类型参数，其指定页面的初始数据、生命周期回调、事件处理函数等。Object 参数说明如表 2-2 所示。

表 2-2 Object 参数说明

属　　性	类　　型	说　　明
data	Object	页面的初始数据
onLoad	function	生命周期回调——监听页面加载
onShow	function	生命周期回调——监听页面显示
onReady	function	生命周期回调——监听页面初次渲染完成
onHide	function	生命周期回调——监听页面隐藏
onUnload	function	生命周期回调——监听页面卸载
onPullDownRefresh	function	监听用户下拉动作
onReachBottom	function	页面上拉触底事件的处理函数
onShareAppMessage	function	用户点击右上角转发
onPageScroll	function	页面滚动触发事件的处理函数
onResize	function	页面尺寸改变时触发
onTabItemTap	function	当前是 Tab 页时，单击 Tab 时触发
其他	any	开发者可以添加任意的函数或数据到 Object 参数中，在页面的函数中用 this 可以访问

Page 函数使用代码如下：

```
Page({
  data: {
    text: "This is page data."
  },
  onLoad: function(options) {
    //页面创建时执行
  },
  onShow: function() {
    //页面出现在前台时执行
  },
  onReady: function() {
    //页面首次渲染完毕时执行
  },
  onHide: function() {
    //页面从前台变为后台时执行
  },
  onUnload: function() {
    //页面销毁时执行
  },
  onPullDownRefresh: function() {
    //触发下拉刷新时执行
  },
  onReachBottom: function() {
    //页面触底时执行
  },
  onShareAppMessage: function () {
    //页面被用户分享时执行
  },
  onPageScroll: function() {
    //页面滚动时执行
  },
  onResize: function() {
    //页面尺寸变化时执行
  },
  onTabItemTap(item) {
    //Tab 单击时执行
    console.log(item.index)
    console.log(item.pagePath)
    console.log(item.text)
  },
  //事件响应函数
  viewTap: function() {
    this.setData({
      text: 'Set some data for updating view.'
    }, function() {
      //this is setData callback
    })
```

```
48        },
49        //自由数据
50        customData: {
51          hi: 'MINA'
52        }
53  })
```

data 是页面第一次渲染使用的初始数据。页面加载时,data 将会以 JSON 字符串的形式由逻辑层传至渲染层,因此 data 中的数据必须是可以转换为 JSON 的类型:字符串、数字、布尔值、对象和数组。

onLoad(Object query):页面加载时触发。一个页面只会调用一次,可以在 onLoad()的参数中获取打开当前页面路径中的参数。

onShow():页面显示/切入前台时触发。

onReady():页面初次渲染完成时触发。一个页面只会调用一次,代表页面已经准备妥当,可以和视图层进行交互。

onHide():页面隐藏/切入后台时触发。如 wx.navigateTo()或底部 Tab 切换到其他页面,微信小程序切入后台等。

onUnload():页面卸载时触发。如 wx.redirectTo()或 wx.navigateBack()到其他页面时。

onPullDownRefresh():监听用户下拉刷新事件。

(1) 需要在 app.json 的 window 选项中或页面配置中开启 enablePullDownRefresh();

(2) 可以通过 wx.startPullDownRefresh()触发下拉刷新,调用后触发下拉刷新动画,效果与用户手动下拉刷新一致;

(3) 当处理完数据刷新后,wx.stopPullDownRefresh()可以停止当前页面的下拉刷新。

onReachBottom():监听用户上拉触底事件。

(1) 可以在 app.json 的 window 选项中或页面配置中设置触发距离 onReachBottomDistance();

(2) 在触发距离内滑动期间,本事件只会被触发一次。

onPageScroll(Object object):监听用户滑动页面事件。

onShareAppMessage(Object object):监听用户单击页面内"转发"按钮(button 组件 open-type="share")或右上角菜单"转发"按钮的行为,并自定义转发内容。注意,只有定义了此事件处理函数,右上角菜单才会显示"转发"按钮。

onResize(Object object):微信小程序屏幕旋转时触发。

2.2.2 页面路由

在微信小程序中所有页面的路由全部由框架进行管理,框架以栈的形式维护了当前的所有页面。当发生路由切换时,页面栈的表现如表 2-3 所示。

表 2-3 发生路由切换时,页面栈的表现

路由方式	页 面 栈 表 现
初始化	新页面入栈
打开新页面	新页面入栈
页面重定向	当前页面出栈,新页面入栈
页面返回	页面不断出栈,直到目标返回页
Tab 切换	页面全部出栈,只留下新的 Tab 页面
重加载	页面全部出栈,只留下新的页面

路由的触发方式以及页面生命周期函数如表 2-4 所示。

表 2-4 路由的触发方式以及页面生命周期函数

路由方式	触发时机	路由前页面	路由后页面
初始化	微信小程序打开的第一个页面		onLoad(),onShow()
打开新页面	调用 API wx.navigateTo() 使用组件 < navigator open-type="navigateTo"/>	onHide()	onLoad(),onShow()
页面重定向	调用 API wx.redirectTo() 使用组件 < navigator open-type="redirectTo"/>	onUnload()	onLoad(),onShow()
页面返回	调用 API wx.navigateBack() 使用组件 < navigator open-type="navigateBack"> 用户单击左上角的"返回"按钮	onUnload()	onShow()
Tab 切换	调用 API wx.switchTab() 使用组件 < navigator open-type="switchTab"/> 用户切换 Tab()		参考表 2-5
重启动	调用 API wx.reLaunch()使用组件 < navigator open-type="reLaunch"/>	onUnload()	onLoad(),onShow()

Tab 切换对应的生命周期(以 A、B 页面为 tabBar 页面,C 是从 A 页面打开的页面,D 页面是从 C 页面打开的页面为例)如表 2-5 所示。

表 2-5 Tab 切换对应的生命周期

当前页面	路由后页面	触发的生命周期(按顺序)
A	A	Nothing happened
A	B	A.onHide(),B.onLoad(),B.onShow()
A	B(再次打开)	A.onHide(),B.onShow()
C	A	C.onUnload(),A.onShow()
C	B	C.onUnload(),B.onLoad(),B.onShow()
D	B	D.onUnload(),C.onUnload(),B.onLoad(),B.onShow()
D(从转发进入)	B	D.onUnload(),B.onLoad(),B.onShow()

这里需要说明路由跳转的几个 API，如表 2-6 所示。

表 2-6　路由的跳转 API

名　　称	功　能　说　明
wx.switchTab	跳转到 tabBar 页面，并关闭其他所有非 tabBar 页面
wx.reLaunch	关闭所有页面，打开到应用内的某个页面
wx.redirectTo	关闭当前页面，跳转到应用内的某个页面
wx.navigateTo	保留当前页面，跳转到应用内的某个页面
wx.navigateBack	关闭当前页面，返回上一页面或多级页面

读者需要了解路由跳转 API 的区别，否则容易造成页面跳转混乱。

注意：

（1）navigateTo()、redirectTo() 只能打开非 tabBar 页面。
（2）switchTab() 只能打开 tabBar 页面。
（3）reLaunch() 可以打开任意页面。
（4）页面底部的 tabBar 由页面决定，即只要是定义为 tabBar 的页面，底部都有 tabBar。
（5）调用页面路由带的参数可以在目标页面的 onLoad() 中获取。

2.2.3　视图层 WXML

WXML（WeiXin Markup Language）是框架设计的一套标签语言，结合基础组件、事件系统，可以构建出页面的结构。WXML 中的动态数据均来自对应 Page 的 data。

以下用一些简单的例子来看 WXML 具有什么功能。

1. 数据绑定

数据绑定使用 Mustache 语法（双大括号）将变量包起来，例如，WXML 代码：

```
1  <view>{{message}}</view>
```

JS 代码：

```
1  Page({
2    data: {
3      message: 'Hello MINA!'
4    }
5  })
```

数据绑定的结果如图 2-8 所示，在页面中显示"Hello MINA!"。

图 2-8　数据绑定的结果

2. 列表渲染

在组件上使用 wx:for 控制属性绑定一个数组，即可使用数组中各项的数据重复渲染该组件。数组当前项的下标变量名默认为 index，数组当前项的变量名默认为 item，例如，WXML 代码：

```
1  < view wx:for = "{{array}}">
2    {{index}}: {{item.message}}
3  </view>
```

JS 代码：

```
1  Page({
2    data: {
3      array: [{
4        message: 'foo',
5      }, {
6        message: 'bar'
7      }]
8    }
9  })
```

列表渲染的结果如图 2-9 所示。

图 2-9 列表渲染的结果

使用 wx:for-item 可以指定数组当前元素的变量名，使用 wx:for-index 可以指定数组当前下标的变量名，例如，WXML 代码：

```
1  < view wx:for = "{{array}}" wx:for - index = "idx" wx:for - item = "itemName">
2    {{idx}}: {{itemName.message}}
3  </view>
```

显示的效果和图 2-9 一致。细心的读者运行上面的列表渲染样例时会发现，Console 窗口中会报一个警告（warning）："Now you can provide attr 'wx:key' for a 'wx:for' to improve performance."意思是说，在使用"wx:for"时你应该使用"wx:key"属性，这样可以提升效率。当数据改变触发渲染层重新渲染时，会校正带有 key 的组件，框架会确保它们被重新排序，而不是重新创建，以确保使组件保持自身的状态，并且提高列表渲染时的效率。

wx:key 的值以两种形式提供：

- 字符串，代表在 for 循环 array 中 item 的某个 property，该 property 的值需要是列表中唯一的字符串或数字，且不能动态改变。
- 保留关键字 * this 代表在 for 循环中的 item 本身，这种表示需要 item 本身是一个唯一的字符串或者数字。

例如，WXML 代码：

```
1  <view wx:for = "{{objectArray}}" wx:key = "unique">{{item.id}}</view>
2  <view>-----------------------------------------------------------</view>
3  <view wx:for = "{{numberArray}}" wx:key = " * this">{{item}}</view>
```

JS 代码：

```
1   Page({
2     data: {
3       objectArray: [
4         {id: 5, unique: 'unique_5'},
5         {id: 4, unique: 'unique_4'},
6         {id: 3, unique: 'unique_3'},
7         {id: 2, unique: 'unique_2'},
8         {id: 1, unique: 'unique_1'},
9         {id: 0, unique: 'unique_0'},
10      ],
11      numberArray: [1, 2, 3, 4]
12    },
13  })
```

图 2-10 演示了 wx:key 值的两种形式使用的效果。

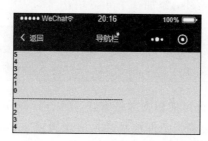

图 2-10 wx:key 值的两种形式使用的效果

当然还可以使用 wx:for 实现嵌套，例如实现一个九九乘法表，WXML 代码：

```
1  <view wx:for = "{{[1, 2, 3, 4, 5, 6, 7, 8, 9]}}" wx:for - item = "i">
2    <view wx:for = "{{[1, 2, 3, 4, 5, 6, 7, 8, 9]}}" wx:for - item = "j">
3      <view wx:if = "{{i <= j}}">
4        {{i}} * {{j}} = {{i * j}}
5      </view>
6    </view>
7  </view>
```

wx:for 嵌套的显示效果如图 2-11 所示。

3. 条件渲染

在框架中，使用 wx:if=""来判断是否需要渲染该代码块：

```
1  <view wx:if = "{{condition}}"> True </view>
```

图 2-11　wx:for 嵌套的显示效果

也可以用 wx:elif 和 wx:else 添加一个 else 块:

```
1  <view wx:if = "{{length > 5}}">1</view>
2  <view wx:elif = "{{length > 2}}">2</view>
3  <view wx:else>3</view>
```

因为 wx:if 之中的模板也可能包含数据绑定,所以当 wx:if 的条件值切换时,框架有一个局部渲染的过程,因为它会确保条件块在切换时销毁或重新渲染。同时 wx:if 也是惰性的,如果在初始渲染条件为 false 时,则框架什么也不做,在条件第一次变成真时才开始局部渲染。相比之下,hidden 就简单得多,组件始终会被渲染,只是简单地控制显示与隐藏。一般来说,wx:if 有更高的切换消耗,而 hidden 有更高的初始渲染消耗。因此,在需要频繁切换的情景下,用 hidden 较好;在运行条件不大可能改变时则用 wx:if 较好。

2.2.4　this.data 和 this.setData 的区别

微信小程序中会经常使用到 this.data 与 this.setData。其中 this.data 是用来获取页面 data 对象的,而 this.setData 是用来更新界面的。那么它们之间有什么区别呢？this.setData 用于将数据从逻辑层发送到视图层(异步),同时改变对应的 this.data 的值(同步)。用 this.data 而不用 this.setData 会造成页面内容不更新的问题。

this.setData 的工作原理如下:微信小程序的视图层使用 WebView 作为渲染载体,而逻辑层是由独立的 JavaScriptCore 作为运行环境。在架构上,WebView 和 JavaScriptCore 都是独立的模块,并不具备数据直接共享的通道。当前,视图层和逻辑层的数据传输,实际上通过两边提供的 evaluateJavaScript 所实现。即用户传输的数据,需要将其转换为字符串形式传递,同时把转换后的数据内容拼接成一份 JS 脚本,再通过执行 JS 脚本的形式传递到两边独立环境。而 evaluateJavaScript 的执行会受很多方面的影响,数据到达视图层并不是实时的。

this.data 可以获取页面 data 对象,但是它返回的对象到底是新的对象还是仅仅只是一个引用?接下来以一个实际的案例来说明 this.data 与 this.setData 的区别。

WXML 代码如下:

```
1  <view wx:for="{{person}}" wx:key="index">
2    {{index}}：{{item}}
3  </view>
4  <button class="cu-btn round bg-green" bindtap="addpersonbyData">使用 this.data 添加人员</button>
5  <button class="cu-btn round bg-green" bindtap="addpersonbySetData">使用 this.setData 添加人员
6  </button>
```

JS 代码如下:

```
1  Page({
2    data: {
3      person: ['塞向秋','瑞访天','乐晏']
4    },
5    onLoad: function (options) {
6      var p = this.data.person
7      p.push('经语心')
8      this.data.person.push('宁曼冬')
9      console.log('onload',p)
10     console.log('onload',this.data.person)
11   },
12   addpersonbyData:function(event)
13   {
14     var p = this.data.person
15     p.push('果明')
16     console.log('addpersonbyData',p)
17     console.log('addpersonbyData',this.data.person)
18   },
19   addpersonbySetData: function (event) {
20     var p = this.data.person
21     p.push('犹淑慧')
22     console.log('addpersonbySetData',p)
23     console.log('addpersonbySetData',this.data.person)
24     this.setData({ person:p })
25   }
26  })
```

页面加载时会执行 onLoad 事件,页面显示效果和 Console 窗口的输出如图 2-12 所示。从图 2-12 中可以看出,在 onLoad 事件中,代码第 7 行改变了 p 变量的值,这时 this.data.

person 的值改变了，代码第 8 行改变了 this.data.person 的值，同样 p 变量的值也改变了。可以看出，p 变量和 this.data.person 变量指向了同一片存储区域，无论通过哪个变量操作这片存储区域，相应的两个变量得出的值都会改变。不过 this.data.person 的值虽然改变了，但是页面上显示的数据却没有变化。

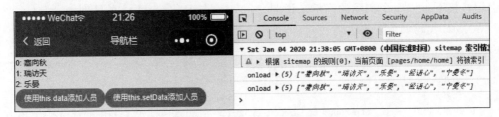

图 2-12　执行 onLoad 事件后页面效果和 Console 窗口输出

接下来单击"使用 this.data 添加人员"按钮，页面显示效果和 Console 窗口的输出如图 2-13 所示。从图 2-13 中看出，onLoad 事件中修改的 this.data.person 的值被保留了下来，和前面一样，虽然修改了 p 变量的值，同样 this.data.person 的值也改变了，但是页面上显示的数据却没有变化。

图 2-13　单击"使用 this.data 添加人员"按钮事件后页面效果和 Console 窗口输出

最后单击"使用 this.setData 添加人员"按钮，页面显示效果和 Console 窗口的输出如图 2-14 所示，这里修改了 p 变量的值，同样 this.data.person 的值也改变了，通过使用 this.setData 页面上的数据发生了改变。

图 2-14　单击"使用 this.setData 添加人员"按钮事件后页面效果和 Console 窗口输出

this.setData 是微信小程序开发中使用最频繁的接口，也是最容易引发性能问题的接口。常见的 this.setData 操作错误如下：

1. 频繁地去 this.setData

部分微信小程序会非常频繁（毫秒级）地去 this.setData，其导致了两个后果：

- Android 下用户在滑动时会感觉到卡顿，操作反馈延迟严重，因为 JS 线程一直在编译执行渲染，未能及时将用户操作事件传递到逻辑层，逻辑层也无法及时将操作处理结果及时传递到视图层。
- 渲染出现延时，但由于 WebView 的 JS 线程一直处于忙碌状态，逻辑层到页面层的通信耗时上升，视图层收到数据消息时距离发出时间已经过去了几百毫秒，渲染的结果并不实时。

2. 每次 this.setData 都传递大量新数据

由 this.setData 的底层实现可知，数据传输实际是一次 evaluateJavaScript 脚本过程，当数据量过大时会增加脚本的编译执行时间，占用 WebView JS 线程。

3. 后台态页面进行 this.setData

当页面进入后台态（用户不可见），不应该继续进行 this.setData，后台态页面的渲染用户是无法感受的，另外后台态页面去 this.setData 也会抢占前台页面的执行。

this.setData 接口的调用涉及逻辑层与渲染层间的线程通信，通信过于频繁可能导致处理队列阻塞，界面渲染不及时而导致卡顿，应避免无用的频繁调用。微信小程序中要求每秒调用 setData 的次数不超过 20 次；由于微信小程序运行逻辑线程与渲染线程之上，this.setData 的调用会把数据从逻辑层传到渲染层，数据太大会增加通信时间，微信小程序中要求 this.setData 的数据在 JSON.stringify 后不超过 256KB。this.setData 操作会引起框架处理一些渲染界面相关的工作，一个未绑定的变量意味着与界面渲染无关，传入 setData 会造成不必要的性能消耗，因此微信小程序中要求 this.setData 传入的所有数据都在模板渲染中有相关依赖。

第3章

云 数 据 库

云数据库提供高性能的数据库写入和查询服务。通过腾讯云开发（Tencent Cloud Base，TCB）的SDK，可以直接在客户端对数据进行读写，也可以在云函数中读写数据，还可以通过控制台对数据进行可视化的增、删、查、改等操作。微信小程序云开发所使用的数据库本质上就是一个MongoDB数据库。MongoDB数据库是介于关系数据库和非关系数据库之间的产品，是非关系数据库中功能最丰富、最像关系数据库的。

数据库：默认情况下，云开发的函数可以使用当前环境对应的数据库。可以根据需要使用不同的数据库。对应MySQL中的数据库。

集合：数据库中多个记录的集合。对应MySQL中的表。

文档：数据库中的一条记录。对应MySQL中的行。

字段：数据库中特定记录的值。对应MySQL中的列。

3.1 云数据库上手

1. 创建第一个集合

打开云开发控制台，选择"数据库"标签页，通过单击集合名称右侧的"＋"按钮创建一个集合。假设要创建一个设备查询类微信小程序，集合名称填写为device。创建成功后，可以看到device集合管理界面，在界面中可以添加记录、查找记录、管理索引和管理权限，如图3-1所示。

2. 创建第一条记录

控制台提供了可视化添加数据的交互界面，选中集合device，单击"添加记录"按钮添加第一条设备数据，如图3-2所示。

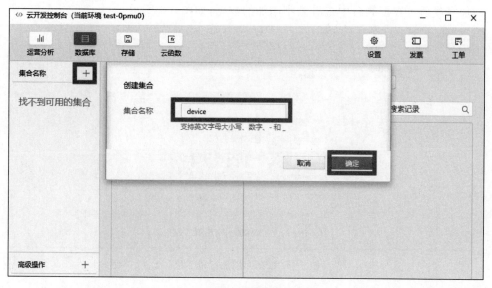

图 3-1 创建集合

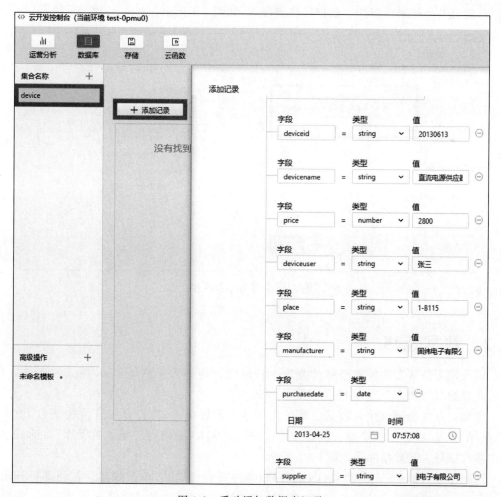

图 3-2 手动添加数据库记录

添加完成后可在控制台中查看到刚添加的数据，如图 3-3 所示。

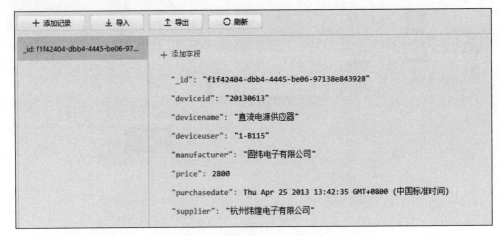

图 3-3　数据库添加记录结果

3. 数据库高级操作

目前云开发控制台新增了数据库高级操作，如图 3-4 所示。

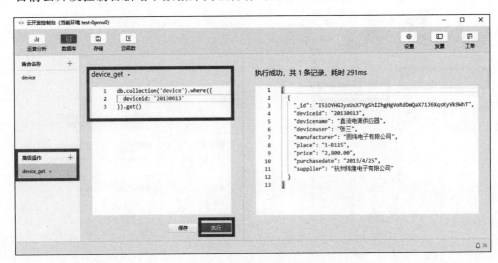

图 3-4　数据库高级操作

3.2　数据迁移

云开发支持从文件导入已有的数据。目前仅支持导入 CSV、JSON 格式的文件数据。有云开发控制台和 HTTP API 两种导入方式。

使用云开发控制台导入数据：要把文件导入云数据库，需打开云开发控制台，切换到"数据库"标签页，并选择要导入数据的集合，单击"导入"按钮，在"导入数据库"对话框中就可以将数据导入云数据库了，如图 3-5 所示。

选择要导入的 CSV 或者 JSON 文件，以及冲突处理模式，单击"确定"按钮即可开始导入。目前云数据库仅支持导入 CSV、JSON 格式的文件数据，有时数据通常以 Excel 的形式

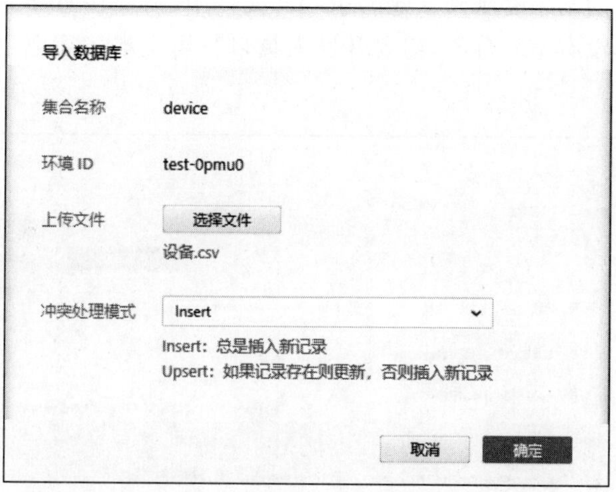

图 3-5 将数据导入云数据库

出现，这里简单介绍 Excel 文件如何转换为 CSV 和 JSON 文件。

Excel 文件转换为 CSV 文件：打开 Excel 文件，然后单击"另存为"，保存类型选择"CSV(逗号分隔)(*.csv)"，因为数据库只支持 UTF-8 格式(默认为 ANSI 编码，导入后中文会显示成乱码)，然后用记事本再次打开 CSV 文件，另存为时选择编码为 UTF-8，如图 3-6 所示，替换原来的 CSV 文件即可。

图 3-6 用 CSV 格式保存为 UTF-8 编码

Excel 文件转换为 JSON 文件：开发者可以专门下载 Excel 转换为 JSON 工具，目前网站也提供了很多在线转换工具，如 http://www.bejson.com/json/col2json/，在线 Excel 文件转换为 JSON 文件的过程如图 3-7 所示。

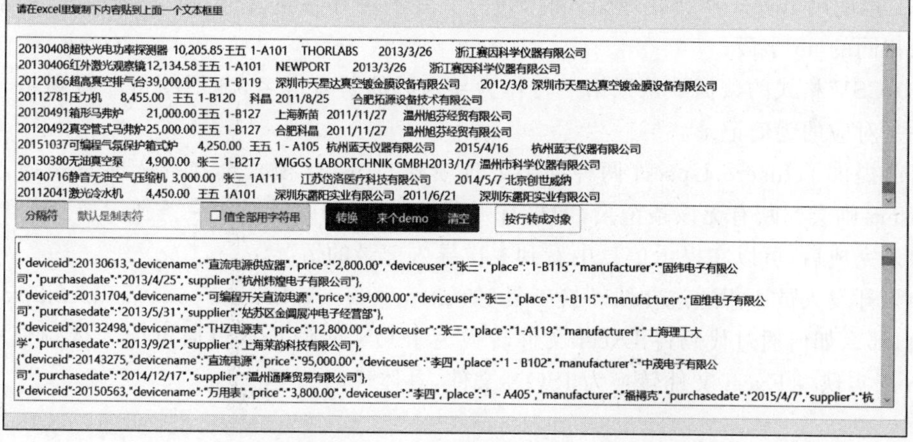

图 3-7 在线 Excel 文件转换为 JSON 文件的过程

然后把在线转换的结果复制到新建的 JSON 文件中,需要注意导入模板的 JSON 格式每条数据记录最后都没有逗号分隔,需要开发人员自行用文本替换把逗号去掉。数据导入结果如图 3-8 所示。

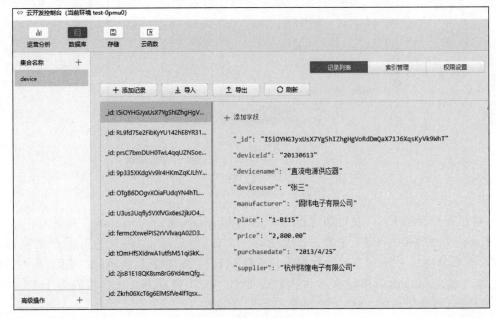

图 3-8　数据导入结果

需要注意以下几点:

(1) JSON 数据不是数组,而是类似 JSON Lines,即各个记录对象之间使用\n 分隔,而非逗号;

(2) JSON 数据每个键值对的键名首尾不能是".",例如".a""abc.",且不能包含多个连续的".",例如"a..b";

(3) 键名不能重复,且不能有歧义,例如{"a":1, "a":2}或{"a": {"b":1},"a.b":2};

(4) 时间格式须为 ISODate 格式,例如"date":{"＄date":"2018-08-31T17:30:00.882Z"};

(5) 当使用 Insert 冲突处理模式时,同一文件不能存在重复的_id 字段,或与数据库已有记录相同的_id 字段;

(6) CSV 格式的数据默认以第一行作为导入后的所有键名,余下的每一行则是与首行键名一一对应的键值记录。

目前提供了 Insert、Upsert 两种冲突处理模式。Insert 模式会在导入时总是插入新记录,Upsert 则会判断有无该条记录,如果有则更新记录,否则就插入一条新记录。

导入完成后,可以在提示信息中看到本次导入记录的情况。

如果开发人员不想通过在线转换工具(如 http://www.bejson.com/json/col2json/)进行转换,那么如何通过代码将 Excel 文件转换为 JSON 文件?本节通过引入依赖模块 xls-to-json 来实现将 Excel 文件转换为 JSON 文件,具体代码如下:

```
1   var node_xj = require("xls-to-json")
2   var fs = require('fs')
3   node_xj({
4     input: "设备.xlsx",              // input xls
5     output: "output.json",           // output json
6     //sheet: "sheet1",               // specific sheetname
7   }, function (err, result) {
8     if (err) {
9       console.error(err)
10    } else {
11      console.log(result)
12      console.log("JSON 文件已经被保存为 output.json.")
13      fs.readFile('output.json', 'utf8', function (err, data) {
14        if (err) {
15          return console.log(err)
16        }
17        var result = data.replace(/},/g, '}\n').replace(/\[{/, '{').replace(/\}]/, '}')
18        fs.writeFile('data.json', result, 'utf8', function (err) {
19          if (err) return console.log(err)
20        })
21      })
22    }
23  });
```

这里使用 VS Code 将本地的设备 Excel 文件转换成 JSON 文件,这里用到 xls-to-json 依赖库,读者使用之前需要在终端使用 npm install xls-to-json 安装此依赖库,代码第 3~7 行将 Excel 文件转换为 JSON 文件,输出文件名为 output.json,但是输出的 JSON 文件数据格式和图 3-7 一致,因此需要进一步修改格式。代码第 13~16 行读取 output.json 文件。代码第 17 行把最前面和最后面的[]去掉,并去掉每条记录后的逗号,最终生成云数据库可直接导入的 JSON 文件为 data.json;读者可以直接把 data.json 文件导入云数据库。

有时记录可能比较复杂,如图 3-9 中的 Excel 数据,其中的选项字段中需要用一个数组来存储,这些数据总不能让用户一条一条地输入数据库,那么这时就需要写代码将 Excel 数据记录转换为 JSON 文件。

针对图 3-9 中的 Excel 数据,Node.js 的代码如下:

```
1   var xlsx = require('node-xlsx')
2   var fs = require('fs')
3   var sheets = xlsx.parse('charpter1.xlsx')
4   var sheet = sheets[0].data
5   var fs = require('fs')
6   var json_data = []
7   for (var rowId = 1; rowId < sheet.length; rowId++) {
8     let data = {}
9     var row = sheet[rowId]
10    data.type = row[0]
11    data.title = row[1]
12    data.option = row[2].split("\r\n");
```

题型	题干	选项	答案	解析	难易度
单选题	计算机所处理的数据一般具有某种内在联系,这是指( )。	A. 数据和数据之间存在某种关系 B. 数据元素和数据元素之间存在某种关系 C. 数据元素内部具有某种结构 D. 数据项和数据项之间存在某种关系	A		易
单选题	数据元素是数据的最小单位。这个断言是( )。	A. 正确的 B. 错误的	A		易
单选题	在数据结构中,与所使用的计算机无关的是数据的( )。	A. 逻辑结构 B. 存储结构 C. 逻辑结构和存储结构 D. 物理结构	A		易
单选题	数据的逻辑结构说明数据元素之间的次序关系,它依赖于数据的存储结构。这个断言是( )。	A. 正确的 B. 错误的	A		易
单选题	数据的存储结构是指数据在计算机内的实际存储形式。这个断言是( )。	A. 正确的 B. 错误的	A		易
单选题	顺序存储方式只能用于线性结构,不能用于非线性结构。这个断言是( )。	A. 正确的 B. 错误的	A		易
单选题	线性结构只能用顺序结构来存放,非线性结构只能用非顺序结构来存放。这个断言是( )。	A. 正确的 B. 错误的	A		易
单选题	计算机算法指的是( )。	A. 计算方法 B. 排序方法 C. 解决问题的有限运算序列 D. 调度方法	A		易
单选题	计算机算法必须具备输入、输出和( )等五个特性。	A. 可行性、可移植性和可扩充性 B. 可行性、确定性和有穷性 C. 确定性、有穷性和稳定性 D. 易读性、稳定性和安全性	A		易
单选题	下面关于算法的说法,正确的是( )。	A. 算法最终必须由计算机程序实现 B. 为解决某问题的算法同与该问题编写的程序含义是相同的 C. 算法的可行性是指指令不能有二义性 D. 以上几个都是错误的	A		易

图 3-9 Excel 数据

```
13      data.answer = row[3]
14      data.explain = row[4]
15      data.level = row[5]
16      json_data.push(data)
17    }
18    var json_str = JSON.stringify(json_data)
19    var result = json_str.replace(/},/g, '}\n').replace(/\[{/, '{').replace(/\}]/, '}')
20    console.log(result)
21    fs.writeFile('result.json', result, 'utf8', function (err) {
22      if (err) return console.log(err);
23    });
```

其中,代码第 3 行采用 node-xlsx 解析 charpter1.xlsx 文件,读者需要安装 node-xlsx 依赖库,在终端中输入 npm install node-xlsx;代码第 7～17 行读取每条记录,将每条记录转行成 JSON 对象,因为第一行对应表头,因此 rowId 从 1 开始,代码第 12 行通过 split("\r\n")方式,把字符串转换为数组;代码第 18 行将 JSON 对象转换为字符串;代码第 19 行把最前面和最后面的[]去掉,并去掉每条记录后的逗号;代码第 21～23 行把 JSON 字符串写入 result.json 文件。然后通过云开发控制台将 result.json 文件导入数据库,如图 3-10 所示。

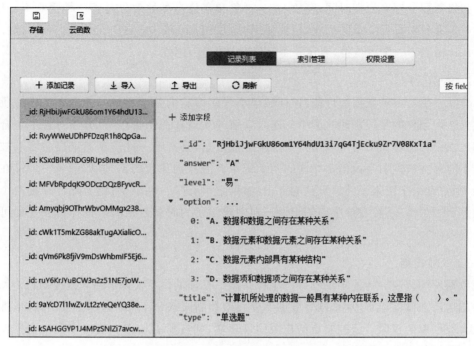

图 3-10　JSON 文件导入数据库

3.3　基础概念

1. 数据类型

云开发数据库提供以下几种数据类型：
- string：字符串。
- number：数字。
- object：对象。
- array：数组。
- bool：布尔值。
- date：时间。
- geo：多种地理位置类型。
- null。

date 类型用于表示时间，精确到毫秒，在微信小程序端可用 JavaScript 内置 Date 对象创建。需要特别注意的是，在微信小程序端创建的时间是客户端时间，不是服务端时间，这意味着在微信小程序端的时间与服务端时间不一定吻合，如果需要使用服务端时间，应该用 API 中提供的 serverDate 对象来创建一个服务端当前时间的标记，当使用了 serverDate 对象的请求抵达服务端处理时，该字段会被转换为服务端当前的时间，更棒的是，在构造 serverDate 对象时还可通过传入一个有 offset 字段的对象来标记一个与当前服务端时间偏移 offset 毫秒的时间，这样可以达到类似如下效果：指定一个字段为服务端时间往后一个小时。

如果需要使用客户端时间,存放 Date 对象和存放毫秒数效果是否一样呢？不一样,因为数据库有针对日期类型的优化,建议使用时都用 Date 或 serverDate 构造时间对象。

null 相当于一个占位符,表示一个字段存在但值为空。

2. 索引管理

建立索引是保证数据库性能、保证微信小程序体验的重要手段。开发人员应为所有需要成为查询条件的字段都建立索引。建立索引的入口在控制台中,可分别对各个集合的字段添加索引。创建索引时可以指定增加唯一性限制,具有唯一性限制的索引会要求被索引集合不能存在被索引字段值都相同的两个记录。需特别注意的是,假如记录中不存在某个字段,则对索引字段来说其值默认为 null,如果索引有唯一性限制,则不允许存在两个或以上的该字段为空/不存在该字段的记录。在创建索引的时候索引属性选择"唯一"即可添加唯一性限制。

3. 权限控制

数据库的权限分为管理端和微信小程序端。管理端包括云函数端和云控制台。微信小程序端运行在微信小程序中,读写数据库受权限控制限制；管理端运行在云函数上,拥有所有读写数据库的权限。云控制台的权限同管理端,拥有所有权限。微信小程序端操作数据库应有严格的安全规则限制。

初期对操作数据库开放以下 4 种权限配置：仅创建者可写,所有人可读；仅创建者可读写；仅管理者可写,所有人可读；仅管理者可读写。每个集合可以拥有一种权限配置,权限配置的规则是作用在集合的每个记录上的。出于易用性和安全性的考虑,云开发为云数据库做了微信小程序深度整合,在微信小程序中创建的每个数据库记录都会带有该记录创建者(即微信小程序用户)的信息,以 _openid 字段保存用户的 openid 在每个相应用户创建的记录中。因此,权限控制也相应围绕一个用户是否应该拥有权限操作其他用户创建的数据展开。

权限级别按照从宽到紧排列如下：

- 仅创建者可写,所有人可读：数据只有创建者可写、所有人可读,如文章。
- 仅创建者可读写：数据只有创建者可读写,其他用户不可读写,如用私密相册。
- 仅管理端可写,所有人可读：该数据只有管理端可写,所有人可读,如商品信息。
- 仅管理端可读写：该数据只有管理端可读写,如后台用的不暴露的数据。

简而言之,管理端始终拥有读写所有数据的权限,微信小程序端始终不能写他人创建的数据。微信小程序端的记录的读写权限其实分为：

(1) 所有人可读,只有创建者可写；

(2) 仅创建者可读写；

(3) 所有人可读,仅管理端可写；

(4) 所有人不可读,仅管理端可读写。

云数据库权限设置如表 3-1 所示。

表 3-1 云数据库权限设置

模　式	微信小程序端读自己创建的数据	微信小程序端写自己创建的数据	微信小程序端读他人创建的数据	微信小程序端写他人创建的数据	管理端读写任意数据
仅创建者可写，所有人可读	√	√	√	×	√
仅创建者可读写	√	√	×	×	√
仅管理端可写，所有人可读	√	×	√	×	√
仅管理端可读写	×	×	×	×	√

集合权限设置如图 3-11 所示。在设置集合权限时应谨慎设置，防止出现越权操作。

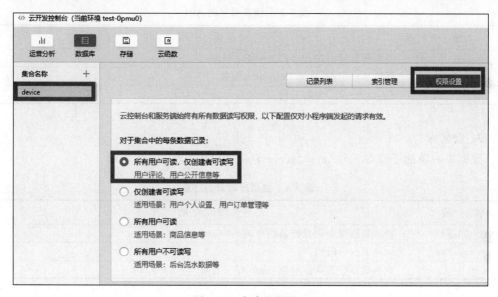

图 3-11　集合权限设置

值得注意的是，新建集合的默认权限是：仅创建者可读写，如果是非创建者在微信小程序端是无法读取集合数据的。有些读者的数据是通过云开发控制台导入的，没有对集合的权限进行修改（默认权限是仅创建者可读写），就会出现在微信小程序读取不到集合数据的情况。

3.4　云数据库 API 列表

微信小程序云开发提供了丰富的数据库操作 API，数据库 API 都是"懒"执行的，这意味着只有真实需要网络请求的 API 调用才会发起网络请求，其余如获取数据库、集合、记录的引用、在集合上构造查询条件等都不会触发网络请求。

1. 请求

请求表示链式调用终结。触发网络请求的 API 有 get、add、update、set、remove 和 count，如表 3-2 所示。

表 3-2 触发网络请求的 API

API	说　明
get	获取集合/记录数据
add	在集合上新增记录
update	更新集合/记录数据
set	替换更新一个记录
remove	删除记录
count	统计查询语句对应的记录条数

2. 引用

获取引用的 API 有 database、collection 和 doc，如表 3-3 所示。

表 3-3 获取引用的 API

API	说　明
database	获取数据库引用，返回 Database 对象
collection	获取集合引用，返回 Collection 对象
doc	获取对一个记录的引用，返回 Document 对象

3. 数据库

数据库对象的字段有 command、serverDate 和 Geo，如表 3-4 所示。

表 3-4 数据库对象的字段

字　段	说　明
command	获取数据库查询及更新指令，返回 Command
serverDate	构造服务端时间
Geo	获取地理位置操作对象，返回 Geo 对象

4. 集合

集合对象 API 有 doc、add、where、orderBy、limit、skip 和 field，如表 3-5 所示。

表 3-5 集合对象 API

API	说　明
doc	获取对一个记录的引用，返回 Document 对象
add	在集合上新增记录
where	构建一个在当前集合上的查询条件，返回 Query，查询条件中可使用查询指令
orderBy	指定查询数据的排序方式
limit	指定返回数据的数量上限
skip	指定查询时从选中的记录列表中的第几项之后开始返回
field	指定返回结果中每条记录应包含的字段

5. 记录/文档

记录/文档对象 API 有 get、update、set、remove 和 field，如表 3-6 所示。

表 3-6　记录/文档对象 API

API	说　　明
get	获取记录数据
update	局部更新数据
set	替换更新记录
remove	删除记录
field	指定返回结果中记录应包含的字段

6. 查询指令

Command 对象查询指令如表 3-7 所示。

表 3-7　Command 对象查询指令

类　　别	指　　令	说　　明
比较	eq	字段是否等于指定值
	neq	字段是否不等于指定值
	lt	字段是否小于指定值
	lte	字段是否小于或等于指定值
	gt	字段是否大于指定值
	gte	字段是否大于或等于指定值
	in	字段值是否在指定数组中
	nin	字段值是否不在指定数组中
逻辑	and	条件与，表示需同时满足多个查询筛选条件
	or	条件或，表示只需满足其中一个条件即可
	nor	表示需所有条件都不满足
	not	条件非，表示对给定条件取反
字段	exists	字段存在
	mod	字段值是否符合给定取模运算
数组	all	数组所有元素是否满足给定条件
	elemMatch	数组是否有一个元素满足所有给定条件
	size	数组长度是否等于给定值
地理位置	geoNear	找出字段值在给定点的附近的记录
	geoWithin	找出字段值在指定区域内的记录
	geoIntersects	找出与给定的地理位置图形相交的记录

7. 更新指令

Command 对象更新指令如表 3-8 所示。

表 3-8　Command 对象更新指令

类　　别	指　　令	说　　明
字段	set	设置字段为指定值
	remove	删除字段
	inc	原子操作，自增字段值
	mul	原子操作，自乘字段值

续表

类别	指令	说明
	min	如果字段值小于给定值,则设为给定值
	max	如果字段值大于给定值,则设为给定值
	rename	字段重命名
数组	push	往数组尾部增加指定值
	pop	从数组尾部删除一个元素
	shift	从数组头部删除一个元素
	unshift	往数组头部增加指定值
	addToSet	原子操作,如果不存在给定元素则添加元素
	pull	剔除数组中所有满足给定条件的元素
	pullAll	剔除数组中所有等于给定值的元素

3.5 云数据库操作

从开发者工具 1.02.1906202 开始,在云控制台数据库管理页中可以编写和执行数据库脚本,脚本可对数据库进行 CRUD 操作,语法同 SDK 数据库语法,如图 3-12 所示。

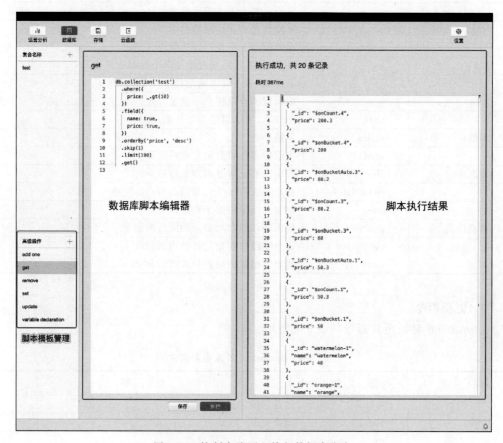

图 3-12 控制台编写和执行数据库脚本

3.5.1 增加记录

可以通过在集合对象上调用 add() 方法往集合中插入一条记录,目前 add() 方法一次只能插入一条记录。添加一条记录的执行语句为:

```
1  db.collection('device').add({
2    data:{
3      "deviceid":"20130410",
4      "devicename":"红外狭缝扫描光束分析仪",
5      "price":42042.29,
6      "deviceuser":"王五",
7      "place":"1－A101",
8      "manufacturer":"THORLABS",
9      "purchasedate":"2013/3/26",
10     "supplier":"浙江赛因科学仪器有限公司"}
11 })
```

在云开发控制台中选择"数据库"标签页,在左侧树形目录中选择"高级操作",执行数据库插入操作,结果如图 3-13 所示。

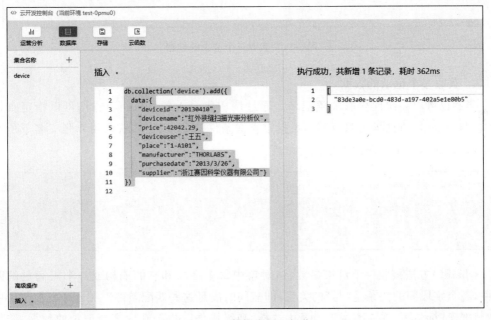

图 3-13　数据库插入操作

3.5.2 查询记录

在记录和集合上都提供了 get() 方法用于获取单个记录或集合中多个记录的数据。

1. 获取一个记录的数据

获取一个记录的数据执行语句为:

```
1  db.collection('device')
2  .doc('muRMVtf5R8RSu74VNS3juEdD7w3IRr2LJWXv6tQU4rZAXYrM')
3  .get()
```

在云开发控制台中选择"数据库"标签页，在左侧树形目录中选择"高级操作"，执行获取一个记录的数据，结果如图 3-14 所示。

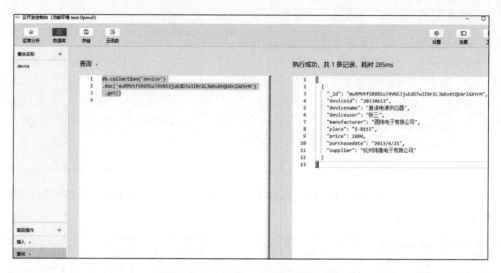

图 3-14 获取一个记录的数据

2. 获取多个记录的数据

当然也可以一次性获取多条记录。通过调用集合上的 where() 方法可以指定查询条件，再调用 get() 方法即可只返回满足指定查询条件的记录，比如获取用户张三名下所有的设备信息，获取多个记录的数据执行语句为：

```
1  db.collection('device').where({
2    deviceuser:"张三"
3  })
4  .get()
```

where() 方法接收一个对象参数，该对象中每个字段和它的值构成一个需满足的匹配条件，各个字段间的关系是"与"的关系，即需同时满足这些匹配条件。在云开发控制台中选择"数据库"标签页，在左侧树形目录中选择"高级操作"，执行获取多个记录的数据，结果如图 3-15 所示。

3. 数据库查询指令

使用数据库 API 提供的 where() 方法可以构造复杂的查询条件完成复杂的查询任务。
API 提供的查询指令，如表 3-9 所示。

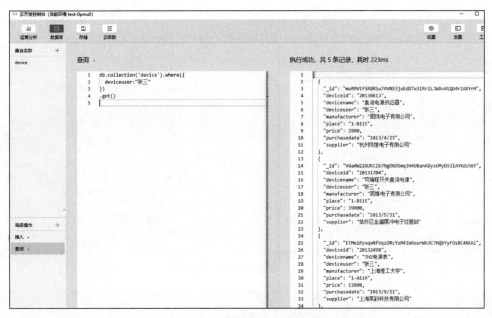

图 3-15 获取多个记录的数据

表 3-9 API 提供的查询指令

指 令	说 明
eq	等于
neq	不等于
lt	小于
lte	小于或等于
gt	大于
gte	大于或等于
in	字段值在给定数组中
nin	字段值不在给定数组中

除了指定一个字段满足一个条件之外,还可以通过指定一个字段需同时满足多个条件,比如用 and 逻辑指令查询价格为 3500~10000 元的设备,执行代码为:

```
1  db.collection('device').where({
2    price:_.gt(3500).and(_.lt(10000))
3  }).get()
```

在云开发控制台中选择"数据库"标签页,在左侧树形目录中选择"高级操作",执行满足多个条件的查询,结果如图 3-16 所示。

既然有 and,当然也有 or 了,Command.or 用于表示逻辑"或"的关系,表示任意满足一个查询筛选条件。"或"指令有两种用法:一是可以进行字段值的"或"操作;二是可以进行跨字段的"或"操作。比如进行字段值的"或"操作,以查询设备为例,查询设备名称为"万用表"或"直流电源"的设备信息,执行代码为:

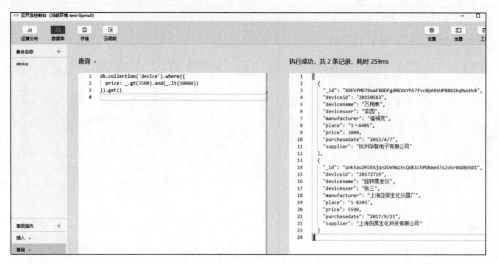

图 3-16　满足多个条件的查询

```
1  db.collection('device').where({
2    devicename: _.eq('万用表').or(_.eq('直流电源'))
3  }).get()
```

在云开发控制台中选择"数据库"标签页，在左侧树形目录中选择"高级操作"，执行字段值"或"操作，结果如图 3-17 所示。

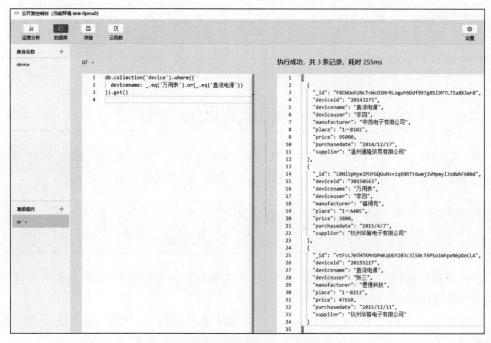

图 3-17　执行字段值"或"操作

还可以进行跨字段的"或"操作，以查询设备为例，查询设备名称为"万用表"或价格为69300 元的设备信息，执行代码为：

```
1   db.collection('device').where(
2     _.or([{
3       devicename: _.eq('万用表')
4     },
5     {
6       price: 69300
7     }
8     ])
9   )
10   .get()
```

在云开发控制台中选择"数据库"标签页,在左侧树形目录中选择"高级操作",执行跨字段"或"操作,结果如图 3-18 所示。

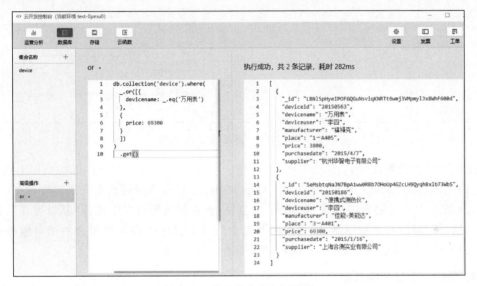

图 3-18　执行跨字段"或"操作

在一个实际的项目中,通常对数据库的操作需要用到集合上多个 API,接下来演示集合上多个 API 如何同时使用。首先用户在云数据库集合 device 中导入 124 条数据记录,用户在云开发控制台高级操作中执行命令:

```
1   db.collection('device')
2     .where({
3       price: _.gt(1000)
4     })
5     .field({
6       deviceid: true,
7       price: true,
8     })
9     .orderBy('price', 'desc')
10    .skip(1)
11    .limit(120)
12    .get()
```

代码第 2～4 行中加入了查询的限制条件；代码第 5～8 行指定返回结果中每条记录应包含的字段；代码第 9 行指定查询数据的排序方式，其中 desc 表示降序，asc 表示升序；代码第 10 行指定查询时从选中的记录列表中的第几项之后开始返回；代码第 11 行指定返回数据的数量上限。在云开发控制台中选择"数据库"标签页，在左侧树形目录中选择"高级操作"，执行集合对象多个 API 查询，结果如图 3-19 所示。

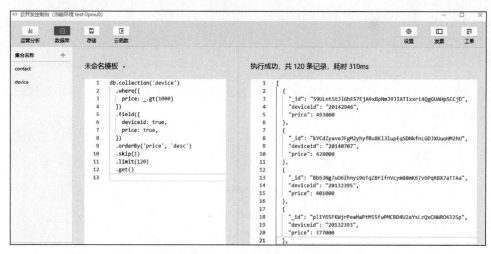

图 3-19　集合对象多个 API 查询

这里有两点需要说明：

（1）API（add、where、orderBy、limit、skip 和 field 等）在调用时没有先后顺序；

（2）在云开发控制台中选择"数据库"标签页，在左侧树形目录中选择"高级操作"，执行查询的返回记录没有数量限制，而在实际开发中微信小程序端在获取集合数据时服务器一次默认并且最多返回 20 条记录，云函数端这个数字则是 100。

4. 获取一个集合的数据

如果要获取一个集合的数据，比如获取 device 集合上的所有记录，可以在集合上调用 get() 方法获取，但通常不建议这么使用，在微信小程序中需要尽量避免一次性获取过量的数据，只应获取必要的数据。开发者可以通过 limit() 方法指定需要获取的记录数量，但微信小程序端不能超过 20 条，云函数端不能超过 100 条。如果要获取集合中所有记录的数据，很可能一个请求无法取出所有数据，需要分批次取。获取一个集合数据的执行代码为：

```
1  const cloud = require('wx-server-sdk')
2  cloud.init()
3  const db = cloud.database()
4  const MAX_LIMIT = 100
5  exports.main = async (event, context) =>{
6    let databasename = event.databasename
7    //先取出集合记录总数
8    const countResult = await db.collection(databasename).count()
9    const total = countResult.total
10   //计算需分几次取
```

```
11    const batchTimes = Math.ceil(total / 100)
12    //承载所有读操作的 promise 的数组
13    const tasks = []
14    for (let i = 0; i < batchTimes; i++) {
15      const promise = db.collection(databasename).skip(i * MAX_LIMIT).limit(MAX_LIMIT).get()
16      tasks.push(promise)
17    }
18    //等待所有
19    return (await Promise.all(tasks)).reduce((acc, cur) =>{
20      return {
21        data: acc.data.concat(cur.data),
22        errMsg: acc.errMsg,
23      }
24    })
25  }
```

async 顾名思义是"异步"的意思，async 用于声明一个函数是异步的。而 await 从字面意思上是"等待"的意思，就是用于等待异步完成。也就是平常所说的异步等待。不过需注意 await 只能在 async 函数中使用。因此在云函数中使用 async-await() 方法，可以把异步请求变为同步请求。代码第 6 行读取的从微信小程序调用云函数传递的数据库集合名称；第 8、9 行读取集合中记录总数；因为云函数读取数据库集合每次最多只能读取 100 条记录，因此代码第 11 行计算需要分几次读取集合中的数据；代码第 13~17 行分批读取集合中的数据。代码第 19~24 行把所有分批读取的数据进行汇总。

云函数的具体使用见第 5 章云函数的介绍，在微信小程序中调用云函数 database 代码如下：

```
1  wx.cloud.callFunction({
2    name: 'database',
3    data: {
4      databasename: "device"
5    }
6  }).then(res =>{
7    console.log(res.result)
8  })
```

获取集合 device 的所有数据，结果如图 3-20 所示，通过云函数 database 把集合 device 中所有的 124 条数据都取出来了。

3.5.3 更新数据

更新数据主要有两个方法，如表 3-10 所示。

图 3-20 获取集合 device 的所有数据

表 3-10 更新数据的方法

API	说明
update	局部更新一个或多个记录
set	替换更新一个记录

1. 局部更新

使用 update() 方法可以局部更新一个记录或一个集合中的记录,局部更新意味着只有指定的字段会得到更新,其他字段不受影响。

局部更新执行代码为:

```
db.collection('device').doc('muRMVtf5R8RSu74VNS3juEdD7w3IRr2LJWXv6tQU4rZAXYrM')
  .update({
    data: {
      price: _.inc(100)
    }
  })
```

上面的示例代码演示了对设备某一条数据执行价格增加 100 的原子操作,用 inc 指令而不是取出值、加 100 再写进去的,其好处在于这个写操作是个原子操作,不会受到并发写的影响,比如同时有两名用户 A 和 B 取了同一个字段值,然后分别加上 100 和 200 再写进数据库,那么这个字段最终结果会是加了 200 而不是 300。如果使用 inc 指令则不会有这个问题。在云开发控制台中选择"数据库"标签页,在左侧树形目录中选择"高级操作",执行局部更新数据,结果如图 3-21 所示。

2. 替换更新

如果需要替换更新一条记录,可以在记录上使用 set() 方法。替换更新意味着用传入的对象替换指定的记录,执行代码为:

```
db.collection('device').doc('muRMVtf5R8RSu74VNS3juEdD7w3IRr2LJWXv6tQU4rZAXYrM')
  .set({
    data: {
      "deviceid":"20130613",
      "devicename":"直流电源供应器",
```

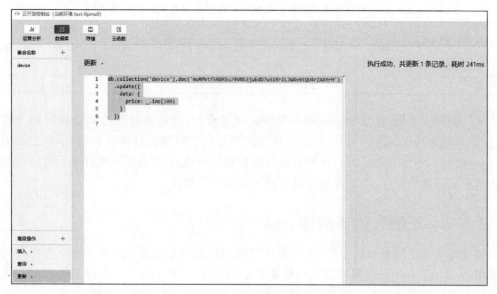

图 3-21 局部更新数据

```
6        "price":2800.00,
7        "deviceuser":"李四",
8        "place":"1 - B105",
9        "manufacturer":"固纬电子有限公司",
10       "purchasedate":"2013/5/25",
11       "supplier":"杭州炜煌电子有限公司"
12    }
13  })
```

使用上面的代码就会把指定的设备信息进行整个替换。

3.5.4 删除数据

对记录使用 remove()方法可以删除该条记录，比如：

```
1  db.collection('device').doc('83de3a0e - bcd0 - 483d - a197 - 402a5e1e80b5')
2    .remove()
```

在云开发控制台中选择"数据库"标签页，在左侧树形目录中选择"高级操作"，执行完上面的数据库删除操作后，会把_id 为'83de3a0e-bcd0-483d-a197-402a5e1e80b5'这条记录进行删除操作。如果需要删除多条记录，则可在 Server 端进行操作（云函数）。

前面介绍了集合中数据的增、删、改、查操作，这些操作是通过在云开发控制台中进行演示，而微信小程序端，云函数端和控制台对数据库操作的权限是不同的，而且能够调用的 API 也是不同的，这里列举微信小程序端和云函数端对数据库调用 API 接口的不同之处，如表 3-11 所示。

表 3-11 微信小程序端和云函数端对数据库调用 API 接口的不同

类别	Collection				Document			
	get	add	update	remove	get	add	update	remove
微信小程序端	√	√	×	×	√	——	√	√
云函数	√	√	√	√	√	——	√	√

从表中可以看出,针对 Document(单条记录操作),微信小程序端和云函数都支持查询、更新和删除操作,针对 Collection(批量记录操作),云函数支持批量查询、单条记录增加(目前一次只能添加一条记录)、批量更新和批量删除操作,但是微信小程序端只支持批量查询和单条记录增加操作,并不支持批量更新和批量删除操作。

3.5.5 正则表达式查询

数据库支持正则表达式查询,开发者可以在查询语句中使用 JavaScript 原生正则对象或使用 db.RegExp()方法来构造正则对象然后进行字符串匹配。在查询条件中对一个字段进行正则匹配即要求该字段的值可以被给定的正则表达式匹配。注意,正则表达式不可用于 db.command 内(如 db.command.in)。

db.RegExp()方法定义如下:

```
1  function RegExp(initOptions: IInitOptions): DBRegExp
2  interface IInitOptions {
3      regexp: string              //正则表达式,字符串形式
4      options: string             //flags,包括 i, m, s,但前端不做强限制
5  }
```

Options 支持 i、m 和 s 这 3 个 flag,注意 JavaScript 原生正则对象构造时仅支持其中的 i 和 m 两个 flag,因此需要使用 s 这个 flag 时必须使用 db.RegExp()构造正则对象。flag 的含义如表 3-12 所示。

表 3-12 flag 的含义

flag	说明
i	大小写不敏感
m	跨行匹配;让开始匹配符^或结束匹配符 $ 除了匹配字符串的开头和结尾外,还匹配行的开头和结尾
s	让. 可以匹配包括换行符在内的所有字符

首先用户在云数据库集合 device 中导入 124 条数据记录,用户可以在 regexp 参数中输入正则表达式(见图 3-22),查询 contact 集合中 mobile 字段以 138 开头、后面 8 个数字结尾的手机号记录,其中^表示匹配输入字行首,\d 表示匹配数字,$ 表示匹配输入行尾。

图 3-23 演示了查询 contact 集合中 name 字段是中文 4 个字的记录,其中^表示匹配输入字行首,$ 表示匹配输入行尾,"\u4e00"和"\u9fa5"是 Unicode 编码,并且正好是中文编码的开始和结束的两个值,所以这个正则表达式可以用来判断 name 包含 4 个中文的记录。

```
db.RegExp ▾                                    执行成功，共 1 条记录，耗时 291ms
1  db.collection('contact').where({           1  [
2      mobile: db.RegExp({                    2    {
3          regexp: '^138\\d{8}$',             3      "_id": "ugXG78jSnQ1cxOhJiIxGGmZQ2GLcIadELqHbV5KR82fDVQ
4          options: 'i',                      4      "mobile": "13858852983",
5      })                                     5      "name": "翁依白",
6  }).get()                                   6      "searchfeild": "wengyibaiwyb翁依白",
7                                             7      "sex": "男"
                                              8    }
                                              9  ]
```

图 3-22　db.RegExp 正则匹配案例 1

```
db.RegExp ▾                                    执行成功，共 1 条记录，耗时 226ms
1  db.collection('contact').where({           1  [
2      name: db.RegExp({                      2    {
3          regexp: '^[\u4e00-\u9fa5]{4}$',    3      "_id": "CetSPX7W3rAiGUTSjVwm4p7P0l8iKQZsTcYmlI2
4          options: 'i',                      4      "mobile": "13165334719",
5      })                                     5      "name": "宇文智刚",
6  }).get()                                   6      "searchfeild": "yuwenzhigangywzg宇文智刚",
7                                             7      "sex": "男"
                                              8    }
                                              9  ]
```

图 3-23　db.RegExp 正则匹配案例 2

图 3-24 演示了查询 contact 集合中 name 字段以白字为结尾的记录，其中 $ 表示匹配输入行尾。

```
db.RegExp ▾                                    执行成功，共 1 条记录，耗时 230ms
1  db.collection('contact').where({           1  [
2      name: db.RegExp({                      2    {
3          regexp: '白$',                     3      "_id": "ugXG78jSnQ1cxOhJiIxGGmZQ2GLcIadELqHbV5K
4          options: 'i',                      4      "mobile": "13858852983",
5      })                                     5      "name": "翁依白",
6  }).get()                                   6      "searchfeild": "wengyibaiwyb翁依白",
7                                             7      "sex": "男"
                                              8    }
                                              9  ]
```

图 3-24　db.RegExp 正则匹配案例 3

上面都是在云开发控制台高级操作中进行演示，接下来以一个实际的通讯录查询案例演示使用正则表达式实现模糊查询。

ColorUI 给出了和手机通讯录类似的页面样式，用微信开发者工具打开 ColorUI 中的 demo 文件，在 tabBar 中选择"扩展"选项，然后单击"索引列表"图标，进入"索引"页面（代码见 pages/plugin/indexes），如图 3-25 所示。

为了便于看到通讯录的效果，本项目提前准备了 JSON 格式的数据（如何从 Excel 文件转换为 JSON 文件，见 3.2 节），导入数据库 contact 集合，导入后的结果如图 3-26 所示，其中 searchfield 字段是手工加进去的，是为了便于对人员信息进行模糊搜索，实现用户在输入框中输入姓名的中文、拼音或者拼音首字母便可以搜索人员。

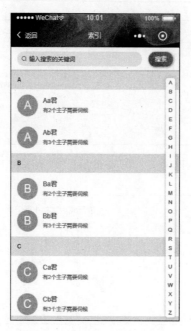

图 3-25 ColorUI 通讯录样式

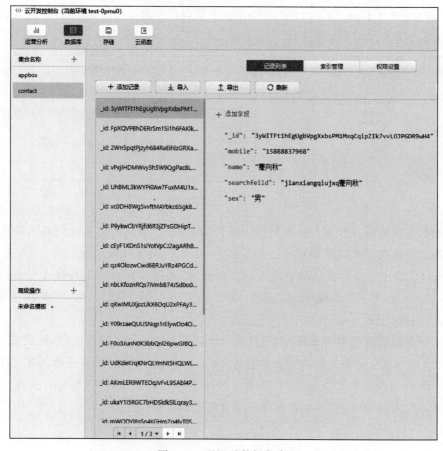

图 3-26 通讯录数据集合

本案例的页面样式 home.wxml 代码如下：

```
1  <cu-custom bgColor = "bg-gradual-pink" isBack = "{{false}}">
2    <view slot = "backText">返回</view>
3    <view slot = "content">通讯录</view>
4  </cu-custom>
5
6  <view class = "cu-bar bg-white search fixed" style = "top:{{CustomBar}}px;">
7    <view class = "search-form round">
8      <text class = "cuIcon-search"></text>
9      <input type = "text" placeholder = "输入名字" confirm-type = "search"
10  <bindinput = "bindKeyInput"></input>
11   </view>
12   <view class = "action">
13     <button class = "cu-btn bg-gradual-green shadow-blur round" bindtap = 'search'>搜索</button>
14   </view>
15  </view>
16  <view class = "cu-list menu-avatar no-padding">
17    <block wx:for = "{{contactCur}}" wx:key wx:for-index = "sub">
18      <view class = "cu-item" bindtap = "showModal" data-id = "{{sub}}" data-target = "bottomModal">
19        <view class = "cu-avatar round lg">{{item.sex}}</view>
20        <view class = "content">
21          <view class = "text-grey">{{item.name}}
22          </view>
23          <view class = "text-gray text-sm">
24            <text class = "text-abc">{{item.mobile}}</text>
25          </view>
26        </view>
27      </view>
28    </block>
29  </view>
```

代码第 6～15 行对应搜索输入框；代码第 16～29 行对应搜索结果显示的样式，样式设计和 ColorUI 基本一致，包括用户性别、用户姓名、用户手机号。

为了显示样式，需要在 home.wxss 中加入如下代码：

```
1  page {
2    padding-top: 100rpx;
3  }
```

相应的 home.js 代码实现如下：

```
1  const app = getApp()
2  const db = wx.cloud.database()
3  Page({
4    data: {
5      StatusBar: app.globalData.StatusBar,
6      CustomBar: app.globalData.CustomBar,
7      contactlist:[]
```

```
 8      },
 9      bindKeyInput: function (event) {
10        const name = event.detail.value
11        db.collection('contact').where({
12          searchfield: db.RegExp({
13            regexp: name,
14            options: 'i',
15          })
16        }).get().then(res =>{
17          this.setData({
18            contactCur: res.data
19          })
20        })
21      }
22    })
```

其中，bindKeyInput 对应输入框输入事件，代码第 11～16 行对应对 searchfield 字段进行模糊查询。对字段进行模糊查询的结果如图 3-27 所示，可以通过姓名的中文、拼音或者拼音首字母进行查询。

图 3-27　进行模糊查询的结果

3.5.6　查询和更新数组元素和嵌套对象

云数据库允许对对象、对象中的元素、数组、数组中的元素进行匹配查询，甚至还可以对数组和对象相互嵌套的字段进行匹配查询/更新，下面从普通匹配开始讲述如何进行匹配查询/更新。

1. 普通匹配

传入的对象的每个<key,value>构成一个筛选条件,有多个<key,value>则表示需同时满足这些条件,是"与"的关系,如果需要"或"关系,可使用[command.or]。

比如找出未完成的进度 50 的待办事项:

```
1  db.collection('todos').where({
2    done: false,
3    progress: 50
4  }).get()
```

2. 匹配记录中的嵌套字段

假设在集合中有如下记录:

```
1  {
2    "style": {
3      "color": "red"
4    }
5  }
```

如果想要找出集合中 style.color 为 red 的记录,那么可以传入相同结构的对象做查询条件或使用"点表示法"查询:

```
1  //方式一
2  db.collection('todos').where({
3    style: {
4      color: 'red'
5    }
6  }).get()
7  //方式二
8  db.collection('todos').where({
9    'style.color': 'red'
10 }).get()
```

3. 匹配数组

假设在集合中有如下记录:

```
1  {
2    "numbers": [10, 20, 30]
3  }
```

可以传入一个完全相同的数组来筛选出这条记录:

```
1  db.collection('todos').where({
2    numbers: [10, 20, 30]
3  }).get()
```

4. 匹配数组中的元素

如果想找出数组字段中数组值包含某个值的记录，可以在匹配数组字段时传入想要匹配的值。如对上面的例子，可传入一个数组中存在的元素来筛选出所有 numbers 字段的值包含 20 的记录：

```
1  db.collection('todos').where({
2    numbers: 20
3  }).get()
```

5. 匹配数组第 n 项元素

如果想找出数组字段中数组的第 n 个元素等于某个值的记录，那么在 < key, value > 匹配中可以以"字段.下标"为 key，目标值为 value 做匹配。如对上面的例子，如果想找出 number 字段第二项的值为 20 的记录，查询如下（注意，数组下标从 0 开始）：

```
1  db.collection('todos').where({
2    'numbers.1': 20
3  }).get()
```

更新也是类似，比如要更新_id 为 test 的记录的 numbers 字段的第二项元素至 30：

```
1  db.collection('todos').doc('test').update({
2    data: {
3      'numbers.1': 30
4    },
5  })
```

6. 结合查询指令进行匹配

在对数组字段进行匹配时，也可以使用如 lt、gt 等指令来筛选出字段数组中存在满足给定比较条件的记录。如对上面的例子，可查找出所有 numbers 字段的数组值中存在包含大于 25 的值的记录：

```
1  const _ = db.command
2  db.collection('todos').where({
3    numbers: _.gt(25)
4  }).get()
```

查询指令也可以通过逻辑指令组合条件，比如找出所有 numbers 数组中存在包含大于 25 的值同时也存在小于 15 的值的记录：

```
1  const _ = db.command
2  db.collection('todos').where({
3    numbers: _.gt(25).and(_.lt(15))
4  }).get()
```

7. 匹配并更新数组中的元素

如果想要匹配并更新数组中的元素,而不是替换整个数组,除了指定数组下标外,还可以:

(1) 更新数组中第一个匹配到的元素。

更新数组字段时可以用字段路径.$的表示法来更新数组字段的第一个满足查询匹配条件的元素。注意,使用这种更新时,查询条件必须包含该数组字段。

假如有如下记录:

```
{
  "_id":"doc1",
  "scores":[10,20,30]
}
{
  "_id":"doc2",
  "scores":[20,20,40]
}
```

让所有 scores 中的第一个 20 的元素更新为 25:

```
//注意:批量更新需在云函数中进行
const _ = db.command
db.collection('todos').where({
  scores: 20
}).update({
  data: {
    'scores.$': 25
  }
})
```

如果记录是对象数组也可以做到,路径如字段路径.$.字段路径。

注意事项:

- 不支持用在数组嵌套数组;
- 如果用 unset 更新操作符,不会从数组中去除该元素,而是置为 null;
- 如果数组元素不是对象,且查询条件用了 neq、not 或 nin,则不能使用.$。

(2) 更新数组中所有匹配的元素。

更新数组字段时可以用字段路径.$[]的表示法来更新数组字段的所有元素。

假如有如下记录:

```
{
  "_id":"doc1",
  "scores": {
    "math":[10,20,30]
  }
}
```

比如让 scores.math 字段中所有数字加 10：

```
1  const _ = db.command
2  db.collection('todos').doc('doc1').update({
3    data: {
4      'scores.math.$[]': _.inc(10)
5    }
6  })
```

更新后 scores.math 数组从[10，20，30]变为[20，30，40]。

如果数组是对象数组也是可以的。假如有如下记录：

```
1  {
2    "_id": "doc1",
3    "scores": {
4      "math": [
5        { "examId": 1, "score": 10 },
6        { "examId": 2, "score": 20 },
7        { "examId": 3, "score": 30 }
8      ]
9    }
10 }
```

可以更新 scores.math 下各个元素的 score 原子自增 10：

```
1  const _ = db.command
2  db.collection('todos').doc('doc1').update({
3    data: {
4      'scores.math.$[].score': _.inc(10)
5    }
6  })
```

8. 匹配多重嵌套的数组和对象

上面所讲述的所有规则都是可以嵌套使用的。假设在集合中有如下记录：

```
1  {
2    "root": {
3      "objects": [
4        {
5          "numbers": [10, 20, 30]
6        },
7        {
8          "numbers": [50, 60, 70]
9        }
10     ]
11   }
12 }
```

下面的查询语句找出集合中所有满足 root.objects 字段数组的第二项的 numbers 字段的第三项等于 70 的记录：

```
1  db.collection('todos').where({
2    'root.objects.1.numbers.2': 70
3  }).get()
```

注意，指定下标不是必需的，例如可以找出集合中所有满足 root.objects 字段数组中任意一项的 numbers 字段包含 30 的记录：

```
1  db.collection('todos').where({
2    'root.objects.numbers': 30
3  }).get()
```

更新操作也是类似，例如要更新 _id 为 test 的 root.objects 字段数组的第二项的 numbers 字段的第三项为 80：

```
1  db.collection('todos').doc('test').update({
2    data: {
3      'root.objects.1.numbers.2': 80
4    },
5  })
```

3.5.7 数据库操作 data 赋值

数据库操作中通常需要设置 data 的值，data 的值本质上是 JSON 格式的对象，JOSN 格式的对象在 2.1.3 节中进行了介绍，JSON 格式的对象必须遵守如下写法：对象被花括号{}包围；对象以键值对书写；键必须是字符串，值必须是有效的 JSON 数据类型；键和值由冒号分隔；每个键值对由逗号分隔。

例如，常见的添加记录的方式为（需要在云数据库添加 device 集合）：

```
1   addData: function(e) {
2     const db = wx.cloud.database()
3     db.collection('device').add({
4       data: {
5         "deviceid": "20130613",
6         "devicename": "直流电源供应器",
7         "price": 2800.00,
8         "deviceuser": "张三",
9         "place": "1-B115",
10        "manufacturer": "固纬电子有限公司",
11        "purchasedate": new Date('2013-4-25'),
12        "supplier": "杭州炜煌电子有限公司",
13        "submitdata": db.serverDate()
14      }
15    })
```

```
16        .then(res =>{
17          console.log(res)
18        })
19  }
```

代码第 7 行 price 字段的类型为数字，第 11 行和第 13 行字段类型为 date 类型，第 13 行获取服务器时间。调用 addData 事件，在数据库中 device 集合中插入一条设备记录，如图 3-28 所示。

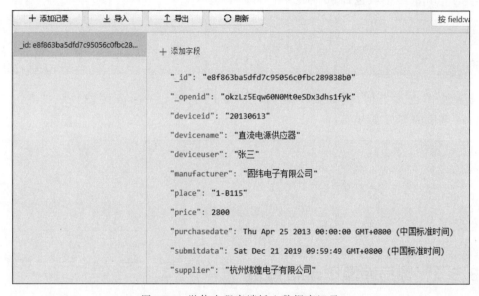

图 3-28　微信小程序端插入数据库记录

有时在插入数据库之前已经有 JSON 对象了，那么是不是也一定需要用键值对的方式写 data 呢？既然 data 的值本质上是 JSON 格式的对象，就可以直接把 JSON 格式的对象赋值给 data，例如：

```
1   addData: function(e) {
2     var device1 = {
3       "deviceid":"20130613",
4       "devicename":"直流电源供应器",
5       "price":2800.00,
6       "deviceuser":"张三",
7       "place":"1-B115",
8       "manufacturer":"固纬电子有限公司",
9       "purchasedate":new Date('2013-4-25'),
10      "supplier":"杭州炜煌电子有限公司"
11    }
12    const db = wx.cloud.database()
13    device1.submitdata = db.serverDate()
14    db.collection('device').add({
15      data: device1
16    })
```

```
17        .then(res =>{
18            console.log(res)
19        })
20  }
```

上面例子实现的结果和图 3-28 一致。

此外，data 数据中的字段还支持对象和数组类型。接下来以数组为例演示如何在 vote 集合中插入一条投票记录(需要在云数据库添加 vote 集合)：

```
1   addData: function (e) {
2       var options = []
3       options[0] = {
4           option: 'A: 800 元以下',
5           num: 0
6       }
7       options[1] = {
8           option: 'B: 800 元至 1000 元',
9           num: 0
10      }
11      options[2] = {
12          option: 'C: 1000 元至 2000 元',
13          num: 0
14      }
15      options[3] = {
16          option: 'D: 2000 元以上',
17          num: 0
18      }
19      const db = wx.cloud.database()
20      db.collection('vote').add({
21          data: {
22              "title":"投票标题:你平均一月用去多少生活费",
23              "type":"单选题",
24              options
25          }
26      })
27      .then(res =>{
28          console.log(res)
29      })
30  }
```

代码第 21～25 行的 data 数据中，插入了 options 数组，插入数据库后该字段的字段名为数组的变量名，插入云数据库后的结果如图 3-29 所示。

前面演示了设置对象字段的值，那么如果需要访问的字段名是一个变量，在 data 中如何进行访问呢？接下来以上面插入的投票记录为例，演示如何更新其中一个字段的值，例如：

```
1   addData: function (e) {
2       var opt = '1'
3       var options = 'options.' + opt + '.num'
4       const db = wx.cloud.database()
5       const _ = db.command
6       console.log(options)
7       db.collection('vote').doc('72527ac65dfd84f1056f295e5b6df49e').update({
8         data: {
9           [options]: _.inc(1)
10        }
11      })
12      .then(console.log)
13      .catch(console.error)
14   }
```

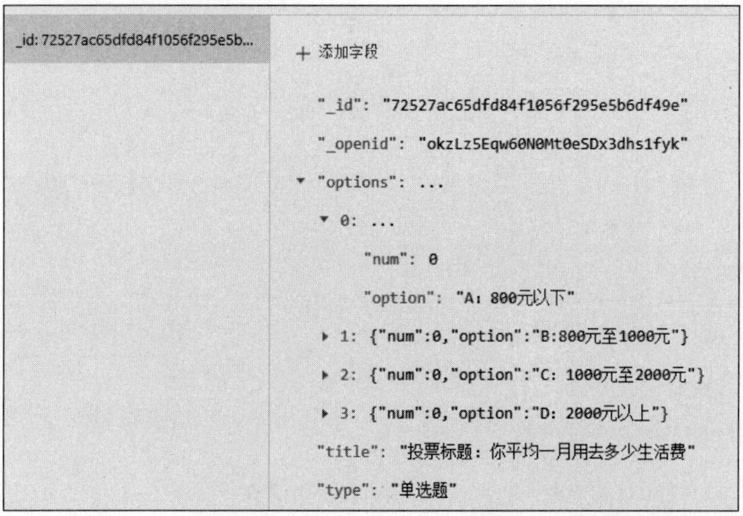

图 3-29　data 字段值为数组

为了演示，代码第 7 行记录的_id 值来源于图 3-29 中的记录，这里假设用户在实际投票页面选择了 B 选项，相应地需要把投票记录中 options[1].num 的值自增 1，在这里相应票数增加采用原子操作 db.command.inc，难点在于票数都在 options 字段中，而 options 本身是个数组，对于更新数组元素的操作读者可以参考 3.5.6 节，这里使用"点表示法"，例如 options.1.num 表示 options 数组中第一个元素中的 num 的值。在这里读者需要注意，data 中的字段名是变量，需要把变量放入[]中，调用 addData 事件以后 vote 集合中相应的选项 num 会增加 1。操作后数据库中的结果如图 3-30 所示。

3.5.8　增、删、改、查案例

接下来以设备数据为例，演示云数据库增、删、改、查的方法。

选择样式：用微信开发者工具打开 ColorUI 中的 demo 文件，在 tabBar 中选择"扩展"选项，然后单击"垂直导航"图标，进入"Tab 索引"页面，ColorUI 中设备查询条目样式如图 3-31 所示。

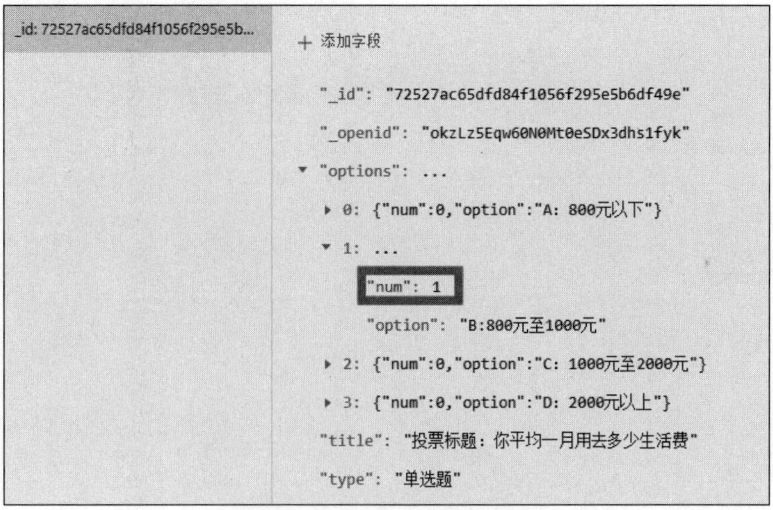

图 3-30　data 字段名是变量

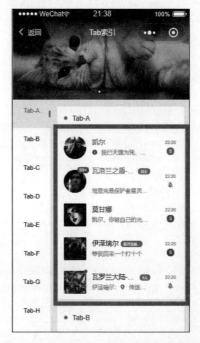

图 3-31　ColorUI 中设备查询条目样式

在 ColorUI 中打开 plugin/verticalnav/verticalnav.wxml 页面找到对应代码，选取如下：

```
1  <view class = "cu-list menu-avatar">
2    <view class = "cu-item">
3      <view class = "cu-avatar round lg"
4  style = " background-image: url ( https://ossweb-img.qq.com/images/lol/web201310/skin/
5  big10001.jpg);"></view>
```

```
6          <view class="content">
7            <view class="text-grey">凯尔</view>
8            <view class="text-gray text-sm flex">
9              <text class="text-cut">
10               <text class="cuIcon-infofill text-red
11 margin-right-xs"></text>我以天理为凭,踏入这片荒芜,不再受凡人的枷锁遏制。我以天理为
12 凭,踏入这片荒芜,不再受凡人的枷锁遏制。
13             </text>
14           </view>
15         </view>
16         <view class="action">
17           <view class="text-grey text-xs">22:20</view>
18           <view class="cu-tag round bg-grey sm">5</view>
19         </view>
20       </view>
21     </view>
```

开发者可以直接使用上面的样式,也可以对上面的样式进行适当的修改,修改后的样式如下(background-image 的图片从 https://www.iconfont.cn 中搜索设备图标下载后,上传至云存储):

```
1  <view class="cu-list menu sm-border margin-top">
2    <block wx:for="{{deviceslist}}" wx:key="index">
3      <view class="cu-item">
4        <view class="cu-avatar round lg"
5  style="background-image:url(https://7465-test-0pmu0-1300559272.tcb.qcloud.la/device/
6  shebei.png?sign=000998c6a7a5ad83e6a677b71d7bf10a&t=1573090290);"></view>
7        <view class="content">
8          <view class="text-grey">
9            <text class="text-lg text-grey">设备名称:</text>
10           <text class="text-lg text-mauve">{{item.devicename}}</text>
11         </view>
12         <view>
13           <text class="text-lg text-grey">设备编号:</text>
14           <text class="text-lg text-mauve">{{item.deviceid}}</text>
15           <text class="text-lg text-grey">领用人:</text>
16           <text class="text-lg text-mauve">{{item.deviceuser}}</text>
17         </view>
18       </view>
19       <view class="action">
20         <button class="cu-btn bg-green margin-tb-sm">编辑</button>
21       </view>
22     </view>
23   </block>
24 </view>
```

在微信小程序页面加载时进行设备查询,进入微信小程序微信官方文档(https://developers.weixin.qq.com/miniprogram/dev/framework/),选择"云开发"子页面,在左侧

树形目录中选择"SDK 文档"→"数据库"→Collection→get 选项,选择数据库集合查询语句,如图 3-32 所示。

```
Promise 风格

const db = wx.cloud.database()
db.collection('todos').where({
  _openid: 'xxx' // 填入当前用户 openid
}).get().then(res => {
  console.log(res.data)
})
```

图 3-32　数据库集合查询语句

复制图 3-32 中的数据库集合查询语句到 home.js 文件中的 onLoad 事件中,并在 data 中加入 deviceslist 对象用于在 home.wxml 中显示,具体如下:

```
1  data: {
2    deviceslist: {},
3  },
4  onLoad: function(options) {
5    const db = wx.cloud.database()
6    db.collection('device').get().then(res =>{
7      //console.log(res.data)
8      this.setData({ deviceslist: res.data })
9    })
10 }
```

采用 Collection.get 获取了多条数据库记录,如图 3-33 所示。

图 3-33　数据库多条数据集合查询结果

接下来演示如何根据条件进行数据查询。首先选择搜索框的样式：用微信开发者工具打开 ColorUI 中的 demo 文件，在 tabBar 中选择"扩展"选项，然后单击"索引列表"图标，进入"索引"页面，选择的搜索框样式如图 3-34 所示。

图 3-34　选择的搜索框样式

相应的代码目录为 plugin/indexes/indexes.wxml，选取代码加入 home.wxml 页面，代码如下：

```
1   <cu-custom bgColor="bg-gradual-pink" isBack="{{false}}">
2     <view slot="backText">返回</view>
3     <view slot="content">设备清单</view>
4   </cu-custom>
5   <view class="cu-bar bg-white search fixed" style="top:{{CustomBar}}px;">
6     <view class="search-form round">
7       <text class="cuIcon-search"></text>
8       <input type="text" placeholder="输入设备编号" bindinput="bindKeyInput"></input>
9     </view>
10    <view class="action">
11      <button class="cu-btn bg-gradual-green shadow-blur round" bindtap='deviceSearch'>搜索</button>
12    </view>
13  </view>
14  <view class="cu-list menu sm-border margin-top" style="padding-top:100rpx;">
15    <block wx:for="{{deviceslist}}" wx:key="index">
16      <view class="cu-item">
17        <view class="cu-avatar round lg"
18  style="background-image:url(https://7465-test-0pmu0-1300559272.tcb.qcloud.la/device/
19  shebei.png?sign=000998c6a7a5ad83e6a677b71d7bf10a&t=1573090290);"></view>
20        <view class="content">
21          <view class="text-grey">
22            <text class="text-lg text-grey">设备名称:</text>
23            <text class="text-lg text-mauve">{{item.devicename}}</text>
24          </view>
25          <view>
26            <text class="text-lg text-grey">设备编号:</text>
27            <text class="text-lg text-mauve">{{item.deviceid}}</text>
28            <text class="text-lg text-grey">领用人:</text>
29            <text class="text-lg text-mauve">{{item.deviceuser}}</text>
30          </view>
31        </view>
32        <view class="action">
33          <button class=" cu-btn bg-green margin-tb-sm" bindtap="recordEdit" data-
34  id="{{item._id}}">编辑</button>
35        </view>
```

```
36      </view>
37    </block>
38  </view>
```

相应地，本案例增加了输入框的输入事件，以及"搜索"按钮的搜索事件，在该事件中，根据用户在输入框输入的设备 ID（deviceid），对云数据库进行数据查询，相应的代码如下：

```
1   const app = getApp();
2   const db = wx.cloud.database()
3   Page({
4     data: {
5       StatusBar: app.globalData.StatusBar,
6       CustomBar: app.globalData.CustomBar,
7       deviceslist: {},
8       deviceid: '',
9     },
10    onLoad: function(options) {
11      db.collection('device').get().then(res =>{
12        this.setData({ deviceslist: res.data })
13      })
14    },
15    bindKeyInput: function (event) {
16      this.setData({
17        deviceid: event.detail.value
18      })
19    },
20    deviceSearch:function(event){
21      var deviceid = this.data.deviceid
22      if (deviceid != '') {
23        db.collection('device').where({
24          deviceid: deviceid
25        }).get().then(res =>{
26          this.setData({ deviceslist: res.data })
27        })
28      }
29      else
30      {
31        db.collection('device').get().then(res =>{
32          this.setData({ deviceslist: res.data })
33        })
34      }
35    },
36    recordEdit:function(event)
37    {
38      wx.navigateTo({
39        url: '../edit/edit?_id=' + event.currentTarget.dataset.id
40      })
41    },
42  })
```

根据条件进行数据查询，页面效果如图3-55所示。代码第20～35行实现了设备记录的查询功能，当输入框中有设备编号时（输入框不为空），对该设备编号进行查询；如果输入框为空，则查找所有设备记录。在这里为了便于演示数据库的操作，查找所有设备记录时，实际上微信小程序只能查找20条数据，读者需要在onReachBottom事件（触底刷新事件）中，使用数据库Collection.skip加载后续的设备记录，该功能在后续章节案例中进行讲解。

接下来演示对单条数据进行更新和删除操作。添加页面pages/edit/edit，选择表单样式。用微信开发者工具打开ColorUI中的demo文件，在tabBar中选择"组件"选项，然后单击"表单"，进入"表单"页面，单条设备数据编辑样式选择如图3-36所示。

图3-35　根据条件进行数据查询

图3-36　单条设备数据编辑样式选择

相应的代码目录为componet/form/form.wxml，选取代码加入edit.wxml页面，代码如下：

```
1   <cu-custom bgColor="bg-gradual-pink" isBack="{{true}}">
2     <view slot="backText">返回</view>
3     <view slot="content">设备编辑</view>
4   </cu-custom>
5
6   <form>
7     <view class="cu-form-group margin-top">
8       <view class="title">
9         <text style="color:red">*</text>设备编号:</view>
10        <input placeholder="请输入设备编号" value='{{deviceinfo.deviceid}}' data-name='deviceid'
```

```
11    bindinput="updateValue"></input>
12    </view>
13    <view class="cu-form-group">
14      <view class="title">
15        <text style="color:red">*</text>设备名称:</view>
16      <input placeholder="请输入设备名称" value='{{deviceinfo.devicename}}' data-name=
17  'devicename' bindinput="updateValue"></input>
18    </view>
19    <view class="cu-form-group">
20      <view class="title">
21        <text style="color:red">*</text>设备领用人:</view>
22      <input placeholder="请输入设备领用人" value='{{deviceinfo.deviceuser}}'
23  data-name='deviceuser' bindinput="updateValue"></input>
24    </view>
25    <view class="cu-form-group">
26      <view class="title">
27        <text style="color:red">*</text>设备产商:</view>
28      <input placeholder="请输入设备产商" value='{{deviceinfo.manufacturer}}'
29  data-name='manufacturer' bindinput="updateValue"></input>
30    </view>
31    <view class="cu-form-group">
32      <view class="title">
33        <text style="color:red">*</text>设备存放地:</view>
34      <input placeholder="请输入设备存放地" value='{{deviceinfo.place}}' data-name='place'
35  bindinput="updateValue"></input>
36    </view>
37    <view class="cu-form-group">
38      <view class="title">
39        <text style="color:red">*</text>设备价格:</view>
40      <input placeholder="请输入设备价格" value='{{deviceinfo.price}}' data-name='price'
41  bindinput="updateValue"></input>
42    </view>
43    <view class="cu-form-group">
44      <view class="title">
45        <text style="color:red">*</text>购买日期:</view>
46      <input placeholder="请输入购买日期" value='{{deviceinfo.purchasedate}}'
47  data-name='purchasedate' bindinput="updateValue"></input>
48    </view>
49    <view class="cu-form-group">
50      <view class="title">
51        <text style="color:red">*</text>供应商:</view>
52      <input placeholder="请输入供应商" value='{{deviceinfo.supplier}}' data-name='supplier'
53  bindinput="updateValue"></input>
54    </view>
55  </form>
56
57  <view class="flex padding justify-center">
58    <button class="cu-btn lg bg-pink" bindtap="deviceupdate">修改设备</button>
59    <button class="cu-btn lg bg-pink" bindtap="deviceDelete">删除设备</button>
60  </view>
```

相应的 edit.js 的完整代码如下：

```js
const db = wx.cloud.database()
Page({
  data: {
    deviceinfo: {}
  },
  onLoad: function(options) {
    const _id = options._id
    db.collection('device').doc(_id).get().then(res =>{
      this.setData({
        deviceinfo: res.data
      })
    })
  },

  updateValue: function(event) {
    let name = event.currentTarget.dataset.name;
    let deviceinfo = this.data.deviceinfo
    deviceinfo[name] = event.detail.value
    this.setData({
      deviceinfo: deviceinfo
    })
  },

  deviceupdate: function(event) {
    let deviceinfo = this.data.deviceinfo
    db.collection('device').doc(deviceinfo._id).set({
      data: {
        deviceid: deviceinfo.deviceid,
        devicename: deviceinfo.devicename,
        deviceuser: deviceinfo.deviceuser,
        manufacturer: deviceinfo.manufacturer,
        place: deviceinfo.place,
        price: parseInt(deviceinfo.price),
        purchasedate: deviceinfo.purchasedate,
        supplier: deviceinfo.supplier
      },
      success: function(res) {
        if (res.stats.updated) {
          console.log("更新成功...")
        }
      }
    })
  },

  deviceDelete: function(event) {
    let deviceinfo = this.data.deviceinfo
    db.collection('device').doc(deviceinfo._id).remove()
      .then(res =>{
```

```
49            console.log("删除成功...")
50            wx.navigateTo({
51              url: '../home/home'
52            })
53          })
54       },
55     })
```

代码第 6~13 行实现页面加载(onLoad 事件)时,根据上一个页面传过来的参数_id,查询数据库中的记录;代码第 15~22 行实现页面中输入框输入事件的监听;代码第 24~43 行实现数据库记录的更新;代码第 45~55 行实现数据库记录的删除操作。

在 home 页面单击设备记录后面的"编辑"按钮,就会进入 edit 页面,edit 页面样式如图 3-37 所示。

图 3-37　edit 页面样式

需要说明的是,本案例在云开发控制台中设置集合 device 的权限设置为"所有用户可读,仅创建者可读写"。由于数据记录是通过导入的方式添加进集合,因此每条数据都没有记录创建者。如果是通过微信小程序添加的记录,就会在每条记录中添加一个_openid 字段,而通过数据导入的方式不会自动添加_openid 字段,因此在 edit 页面中直接单击"修改设备"按钮,会提示类似的错误:

```
1  Error: errCode: - 502001 database request fail | errMsg: [FailedOperation.set] multiple
2  write errors: [{write errors: [{E11000 duplicate key error collection: tnt - 12p3936xo.x -
3  j-1 index: id dup key: { : "xjl" }}]}, {<nil>}]
```

造成上面这种错误的主要原因是这条数据不是用户自己创建的,因为集合 device 只支持创建者可读写。要解决这个问题,有两种方法:

(1) 在每条记录中添加_openid 字段,值为开发者自己的 openid,但是该方法也仅支持创建者对该记录的写操作;

(2) 通过云函数,所有用户都可对该记录进行读写。

在这里通过添加_openid 字段的方式,如图 3-38 所示。

图 3-38　数据添加_openid 字段

_openid 的值可以通过云函数获得,也可以进入云开发控制台,选择"运营分析"→"用户访问"选项,直接复制 openid。

第 4 章

云 存 储

云存储提供高扩展性、低成本、可靠和安全的文件存储服务。开发者可以快速地实现文件上传、下载和对应的管理功能。存储支持灵活的鉴权策略,可以满足开发者不同场景下的文件访问管理。开发者可以在客户端和云函数端通过 API 直接使用云存储。

4.1 管理文件

登录云开发控制台,选择"存储"标签页,进入云开发控制台云存储页面,如图 4-1 所示。在此页面可以查看云存储空间中所有的文件。

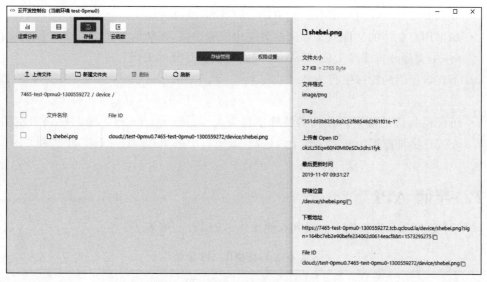

图 4-1 云开发控制台云存储页面

如需查看某个文件详细信息,单击该记录文件名称字段,即可查看关于此文件的所有信息（文件名称、文件大小、存储位置等）。用户可以在该页面新建文件夹；也可以通过单击"上传文件"按钮上传文件，通过单击"删除"按钮删除勾选的文件。

如需设置文件权限，选择"权限设置"，进入云存储权限设置页面，如图 4-2 所示。可根据实际需求，选择相应的权限，默认为"所有用户可读，仅创建者可读写"。

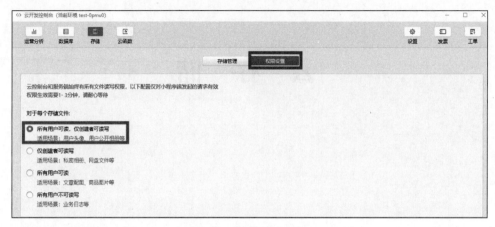

图 4-2　云存储权限设置

文件名命名限制：
- 不能为空；
- 不能以/开头；
- 不能出现连续/；
- 编码长度最大为 850 字节；
- 推荐使用大小写英文字母、数字，即[a~z,A~Z,0~9]和符号-、!、_、.、* 及其组合；
- 不支持 ASCII 控制字符中的字符上(↑)、字符下(↓)、字符右(→)、字符左(←)，其分别对应 CAN(24)、EM(25)、SUB(26)、ESC(27)；
- 如果用户上传的文件或文件夹的名字中带有中文，在访问和请求这个文件或文件夹时，中文部分将按照 URL Encode 规则转换为百分号编码；
- 不建议使用的特殊字符为 '、^、"、\、{、}、[、]、~、%、#、>、<及 ASCII 128~255 十进制；
- 可能需特殊处理后再使用的特殊字符为，、:、;、=、&、$、@、+、?（空格）及 ASCII 字符范围为 00~1F，十六进制(0~31,十进制)以及 7F(127 十进制)。

4.2　存储 API

微信小程序云开发提供了一系列存储操作 API 指令，如表 4-1 所示。

表 4-1　存储操作 API 指令

指　　令	说　　明
uploadFile	上传文件
downloadFile	下载文件

续表

指　　令	说　　明
deleteFile	删除文件
getTempFileURL	换取临时链接

4.3　存储操作

1. 上传文件

在微信小程序端可调用 wx.cloud.uploadFile()方法上传文件:

```
1  wx.cloud.uploadFile({
2    cloudPath: 'example.png',
3    filePath: '',                //文件路径
4  }).then(res => {
5    // get resource ID
6    console.log(res.fileID)
7  }).catch(error => {
8    // handle error
9  })
```

文件上传成功后会获得文件唯一标识符,即文件 ID,后续操作都基于文件 ID 而不是 URL。

2. 下载文件

可以根据文件 ID 下载文件,用户仅可下载其有访问权限的文件:

```
1  wx.cloud.downloadFile({
2    fileID: 'a7xzcb'
3  }).then(res => {
4    // get temp file path
5    console.log(res.tempFilePath)
6  }).catch(error => {
7    // handle error
8  })
```

3. 删除文件

可以通过 wx.cloud.deleteFile 删除文件:

```
1  wx.cloud.deleteFile({
2    fileList: ['a7xzcb']
3  }).then(res => {
4    // handle success
5    console.log(res.fileList)
6  }).catch(error => {
7    // handle error
8  })
```

4. 换取临时链接

可以根据文件 ID 换取临时文件链接,文件链接有效期为两小时:

```
1  wx.cloud.getTempFileURL({
2    fileList: [{
3      fileID: 'a7xzcb',
4      maxAge: 60 * 60, // one hour
5    }]
6  }).then(res =>{
7    // get temp file URL
8    console.log(res.fileList)
9  }).catch(error =>{
10   // handle error
11 })
```

4.4 云存储案例

接下来演示在微信小程序中如何使用云存储,实现文件的上传、下载、删除操作。选择样式:用微信开发者工具打开 ColorUI 中的 demo 文件,在 tabBar 中选择"组件",然后单击"表单",进入"表单"页面,选择的文件上传组件样式(代码见 pages/component/form),如图 4-3 所示。

图 4-3　文件上传组件样式

这里主要用到了图 4-3 中图片上传组件,在本案例中,该组件用来进行文件上传,且只允许用户上传一个文件。在此样式上进行适当修改,在微信小程序中添加文件页面样式代码(pages/home/home.wxml 页面)如下:

```
1  <cu-custom bgColor="bg-gradual-pink" isBack="{{false}}">
2    <view slot="backText">返回</view>
3    <view slot="content">添加文件</view>
4  </cu-custom>
5
```

```
6   <form>
7     <view class = "cu-form-group margin-top">
8       <view class = "title">
9         <text style = "color:red">*</text>文件名称:</view>
10      <input bindinput = "updateValue" data-name = 'title' placeholder = "请输入公告名称"></input>
11    </view>
12    <view class = "cu-form-group">
13      <view class = "title">
14        <text style = "color:red">*</text>发布人:</view>
15      <input bindinput = "updateValue" data-name = 'publisher' placeholder = "请输入发布人"></input>
16    </view>
17    <view class = "cu-form-group align-start">
18      <view class = "title">
19        <text style = "color:red">*</text>文件简介:</view>
20      <textarea maxlength = "-1" bindinput = "updateValue" data-name = 'content' placeholder = "输
21  入公告简介"></textarea>
22    </view>
23    <view class = "cu-form-group">
24      <view class = "title">
25        是否置顶:</view>
26      <switch class = "orange radius sm" bindchange = "switchChange"></switch>
27    </view>
28    <view class = "cu-bar bg-white margin-top">
29      <view class = "action">
30        文件上传
31      </view>
32      <view class = "action">
33        {{fileinfo.fileImg.length}}/1
34      </view>
35    </view>
36    <view class = "cu-form-group">
37      <view class = "grid col-3 grid-square flex-sub">
38        <view class = "bg-img" wx:for = "{{fileinfo.fileImg}}" wx:key = "{{index}}"
39  data-url = "{{fileinfo.fileImg[index]}}">
40          <image src = '{{fileinfo.fileImg[index]}}' mode = 'aspectFill'></image>
41          <view class = "cu-tag bg-red" catchtap = "deleteFile" data-id = "{{index}}">
42            <text class = "cuIcon-close"></text>
43          </view>
44        </view>
45        <view class = "solids" bindtap = "chooseFile" wx:if = "{{fileinfo.fileImg.length<1}}">
46          <text class = "cuIcon-file"></text>
47        </view>
48      </view>
49    </view>
50    <text>{{fileinfo.filename}}</text>
51  </form>
52  
53  <view class = "flex padding justify-center">
54    <button class = "cu-btn lg bg-pink" bindtap = "submitform">提交</button>
55  </view>
```

相应的 pages/home/home.js 代码如下：

```js
const db = wx.cloud.database()
Page({
  data: {
    fileinfo: { istop: 0, fileImg:[]}
  },
  updateValue: function (event) {
    let name = event.currentTarget.dataset.name;
    let fileinfo = this.data.fileinfo
    fileinfo[name] = event.detail.value
    this.setData({
      fileinfo: fileinfo
    })
  },
  switchChange: function (e) {
    let fileinfo = this.data.fileinfo
    if (e.detail.value == true) {
      fileinfo.istop = 1
    }
    else {
      fileinfo.istop = 0
    }
    this.setData({
      fileinfo: fileinfo
    })
  },
  //上传文件
  chooseFile() {
    var that = this
    let fileinfo = this.data.fileinfo
    wx.chooseMessageFile({
      count: 1,
      type: 'file',
      success(res) {
        wx.showLoading({
          title: '文件上传中...',
        })
        //产生 6 位随机数,防止文件名冲突
        var randString = Math.floor(Math.random() * 1000000).toString()
        var filepath = randString + filename
        //文件上传至云存储
        wx.cloud.uploadFile({
          cloudPath: 'file/' + filepath,
          filePath: res.tempFiles[0].path,
        }).then(res =>{
          var filename1 = filename;
          var index1 = filename1.lastIndexOf(".");
          var index2 = filename1.length;
          var type = filename1.substring(index1, index2);
```

```javascript
            var FileImg = []
            if (type == '.pdf') {
                FileImg[0] = '/images/pdf.png'
            }
            else if (type == '.doc' || type == '.docx') {
                FileImg[0] = '/images/word.png'
            }
            else if (type == '.xls' || type == '.xlsx') {
                FileImg[0] = '/images/excel.png'
            }
            else
            {
                FileImg[0] = '/images/file.png'
            }
            fileinfo.fileID = res.fileID
            fileinfo.filename = filename
            fileinfo.fileImg = FileImg
            that.setData({ fileinfo: fileinfo })
            wx.hideLoading()
        }).catch(error =>{
            // handle error
        })
      }
    })
  },
  //删除文件
  deleteFile: function (event) {
    let fileinfo = this.data.fileinfo
    var fileID = fileinfo.fileID
    console.log(fileID)
    wx.cloud.deleteFile({
      fileList: [fileID]
    }).then(res =>{
      fileinfo.fileImg = []
      fileinfo.filename = ''
      this.setData({ fileinfo: fileinfo})
    }).catch(error =>{
      // handle error
    })
  },
  //提交表单
  submitform: function (event) {
    let fileinfo = this.data.fileinfo
    db.collection('files').add({
      //data 字段表示需新增的 JSON 数据
      data: {
        title: fileinfo.title,
        content: fileinfo.content,
        publisher: fileinfo.publisher,
        fileID: fileinfo.fileID,
```

```
99                fileImg: fileinfo.fileImg,
100               filename: fileinfo.filename,
101               istop: fileinfo.istop,
102               submitdate: db.serverDate()
103             }
104           })
105           .then(res =>{
106             wx.showToast({
107               title: '文件添加成功',
108               icon: 'success',
109               duration: 2000
110             })
111             wx.navigateTo({
112               url: '../filelist/filelist'
113             })
114           })
115           .catch(console.error)
116       },
117     })
```

代码第 27～73 行实现用户选择文件进行上传，文件上传至云存储，wx.chooseMessageFile() 实现从客户端选择文件，为了防止文件名上传至云存储后出现文件名冲突，本案例中在文件名之前增加了 6 位随机数，用户也可以在文件名之前加上上传文件的时间。wx.cloud.uploadFile()实现文件上传至云存储，为了区别上传文件的类别，在本案例中把上传的文件分为 Word 文档、PDF 文档、Excel 文档和其他，并分别给了不同的文件图标，便于后面的文件显示。代码第 75～88 行实现了云存储文件的删除功能，其中 wx.cloud.deleteFile()实现云文件删除操作。代码第 90～117 行实现表单提交，把文件的记录添加进入数据库，这里为了简化，并未对用户提交的每个字段的有效性进行检查。代码第 102 行，在该记录中添加了用户提交的服务器时间。

"添加文件"页面显示效果如图 4-4 所示。数据库集合 files 包括如下字段：文件名称、发布人、文件简介、是否置顶、文件云存储地址、文件上传日期等，如图 4-5 所示。用户需要提前建立数据库集合 files，并设置权限：所有用户可读，仅创建者可读写。

然后新增"显示文件"页面 pages/filelist/filelist，其中 filelist.wxml 代码如下：

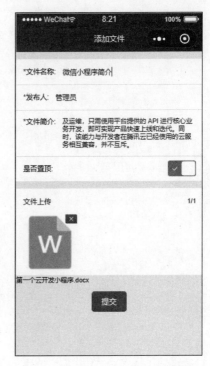

图 4-4 "添加文件"页面显示效果

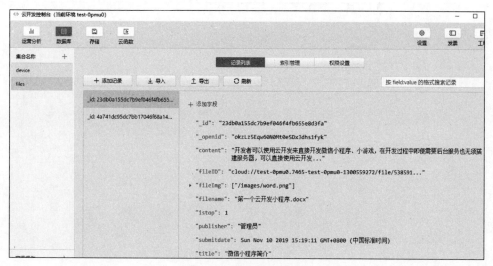

图 4-5 集合 files 字段

```
1   <text>pages/filelist/filelist.wxml</text>
2   <cu-custom bgColor="bg-gradual-pink" isBack="{{true}}">
3     <view slot="backText">返回</view>
4     <view slot="content">显示文件</view>
5   </cu-custom>
6
7   <view class="cu-list menu sm-border">
8     <block wx:for="{{fileslist}}" wx:key="index">
9       <view class="cu-item">
10        <image style="width:80rpx; height:80rpx" src="{{item.fileImg}}" mode="aspectFill"></image>
11        <view class="content">
12          <view class="text-df text-blue">{{item.filename}}</view>
13          <view class="text-gray text-sm">
14            {{item.publisher}}{{item.time}}
15          </view>
16        </view>
17        <view class="action">
18          <button class="cu-btn round bg-green" bindtap="downloadFile" data-id=
19  "{{item.fileID}}">下载</button>
20        </view>
21      </view>
22    </block>
23  </view>
24
```

相应的 filelist.js 代码如下：

```
1   const db = wx.cloud.database()
2   var util = require('../../utils/util.js')
3   Page({
4     data: {
```

```
5        fileslist: {}
6      },
7
8      onLoad: function(options) {
9        db.collection('files').orderBy('istop', 'desc').orderBy('submitdate', 'desc').get().then(res =>{
10         for (var index in res.data) {
11           res.data[index].time = util.formatTime(res.data[index].submitdate)
12         }
13         this.setData({
14           fileslist: res.data
15         })
16       })
17     },
18
19     downloadFile: function(event) {
20       wx.showLoading({
21         title: '文档打开中...',
22       })
23       wx.cloud.downloadFile({
24         fileID: event.currentTarget.dataset.id
25       }).then(res =>{
26         console.log(res.tempFilePath)
27         const filePath = res.tempFilePath
28         wx.openDocument({
29           filePath: filePath,
30           success: res =>{
31             console.log('打开文档成功')
32             wx.hideLoading()
33           }
34         })
35       }).catch(error =>{
36         // handle error
37       })
38     },
39   })
```

代码第 8～17 行在 onLoad 页面数据加载事件中，查找数据库中集合 files 中所有的文件记录，并按照是否置顶排序，随后按照文件上传时间进行排序。代码第 19～38 行从云存储中下载指定的文件，并进行浏览，其中 wx.cloud.downloadFile() 实现下载文件操作，wx.openDocument() 实现打开文档进行浏览操作。在代码第 11 行中，由于上传时间数据类型为 object，util.formatTime() 把 object 类型的时间转换成 string 类型的数据进行显示。根目录下 util.js 文件中的代码如下：

```
1  const formatTime = date =>{
2    const year = date.getFullYear()
3    const month = date.getMonth() + 1
4    const day = date.getDate()
5    const hour = date.getHours()
```

```
6      const minute = date.getMinutes()
7      const second = date.getSeconds()
8      return [year, month, day].map(formatNumber).join('/') + ' ' + [hour, minute, second].map
9    (formatNumber).join(':')
10   }
11   const formatNumber = n =>{
12     n = n.toString()
13     return n[1] ? n : '0' + n
14   }
15   module.exports = {
16     formatTime: formatTime
17   }
```

filelist 页面的显示效果如图 4-6 所示。用户单击"下载"按钮，可从云存储中下载文件，并打开文档进行查看。

图 4-6 filelist 页面的显示效果

第 5 章

云 函 数

云函数是一段运行在云端的代码,无须管理服务器,在开发工具内编写、一键上传部署即可运行后端代码。云开发中的云函数可让用户将自身的业务逻辑代码上传,并通过云开发的调用触发函数,从而实现后端的业务运作。用户可在客户端直接调用云函数,也可以在云函数之间实现相互调用,云函数现在唯一支持的语言为 Node.js 8.9。

5.1 云函数发送 HTTP 请求

在微信小程序和云函数中都可以发送 HTTP 请求,那么什么时候要用云函数发送请求呢?微信对微信小程序端发送 HTTP 请求做了限制:

(1)微信小程序端使用 wx.request 可以发起一个 HTTP 请求,一个微信小程序被限制为同时只有 5 个网络请求,而使用云函数发送 HTTP 请求没有这个限制;

(2)微信小程序端发送 HTTP 请求域名必须经过 ICP 备案,而云函数不需要进行域名备案;

(3)微信小程序端发送请求域名只支持 HTTPS,且要在微信公众平台进行服务器域名配置,而云函数支持 HTTP/HTTPS,不需要进行服务器域名配置。

如果开发者自己开发后端并搭建服务器,且在开发阶段还没有进行域名备案,可以设置不校验域名和 HTTPS 证书,如图 5-1 所示,在微信开发者工具中查看,单击"详情"按钮,然后选择"本地设置"子页面,勾选"不校验合法域名、web-view(业务域名)、TLS 版本以及 HTTPS 证书"复选框。

本节介绍在云函数中使用 got 发送 HTTP 请求,got 是为 Node.js 获取人性化且功能强大的 HTTP 请求库,而云函数使用的就是 Node.js。接下来演示如何在云函数中使用 got 发送 HTTP 请求。

1. 在云函数中安装 got 依赖库

在微信开发者工具中,右击目录 cloudfunctions,在弹出的快捷菜单中选择"新建 Node.js 云函数"选项,如图 5-2 所示,输入云函数名称,比如 HTTP,这样一个云函数就建好了。

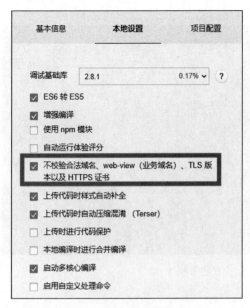

图 5-1　设置不校验域名和 HTTPS 证书

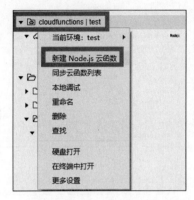

图 5-2　新建 Node.js 云函数

新建的 Node.js 云函数其实是存在于本地目录中的,在本地目录中编写完云函数,再把云函数上传到云端。上传云端的方法有两种:"上传并部署:云端安装依赖(不上传 node_modules)"和"上传并部署:所有文件"。第一种方法只会上传本地的代码,安装的依赖库不会上传;第二种方法是把本地的代码和依赖库一同上传至云端。那么如果采用第一种方法,云端是怎么知道要安装哪个依赖库?答案是云端会根据云函数 package.json 文件中的 dependencies 安装相应的依赖库。云函数除了在本地调试时需要在本地目录安装依赖库以外,其他都是在云端调用的,因此通常情况下只要安装云端的依赖库就可以了,本地目录不需要安装依赖库。

根据前面的介绍,在云端安装依赖库,只需要在本地目录云函数 package.json 文件中的 dependencies 写入要安装的库就可以了,有两种方法实现:

方法一:通过 npm 安装依赖库,比如安装 got 依赖库,如图 5-3 所示。

```
npm install got
```

安装方法为:右击云函数目录,在弹出的快捷菜单中选择"在终端中打开"选项,随后在打开的控制台中输入 npm install got。该方法会在本地目录安装 got,同时写入 package.json 文件中的 dependencies。

方法二:开发者直接手动在 package.json 文件中的 dependencies 中写入需要安装的依赖库。如果不知道依赖库的版本,则可以直接写"latest",表示最新版,比如""dependencies": {"got":"^9.6.0"}",如图 5-4 所示。该方法不会在本地目录安装依赖库。需要说明的是,作者在写本书时最新版本 got 库不支持 Node.js 8.9,因此读者如果需要使用 got 库,则可以使用上面的 9.6.0 版本。

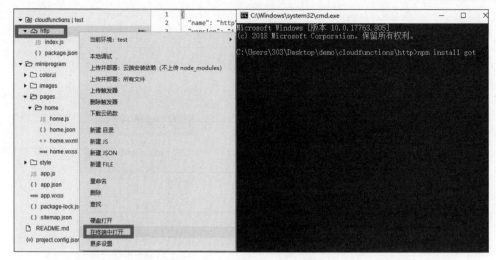

图 5-3 通过 npm 安装依赖库

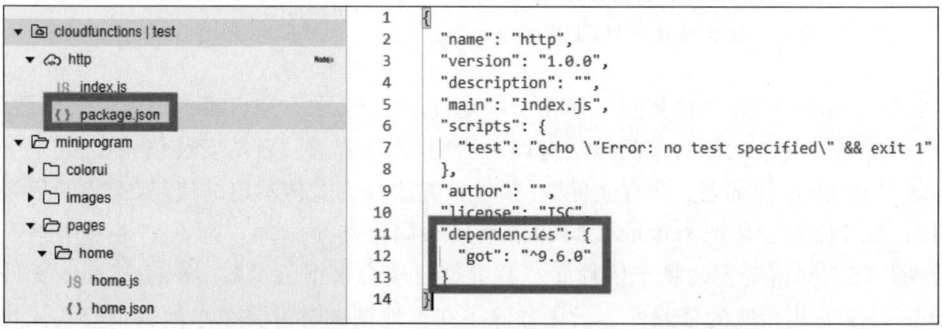

图 5-4 package.json 文件中的 dependencies 依赖库

方法一通过 npm 会在本地安装依赖库,方法二则不会在本地安装依赖库,两种方法都会在 package.json 文件中的 dependencies 中写入需要安装的依赖库。

在本地配置好云函数的依赖库后,右击云函数目录,在弹出的快捷菜单中选择"上传并部署:云端安装依赖(不上传 node_modules)"选项,云端会根据 package.json 文件中的 dependencies 自动安装依赖库。如果用户不需要在本地调试云函数,则建议直接采用方法二。

2. 在云函数中使用 got 发送 HTTP 请求

在云函数中使用 got 发送 GET 请求代码如下:

```
const got = require('got');
exports.main = async (event, context) =>{
  let getResponse = await got('httpbin.org/get')
  return getResponse.body
}
```

在云函数中使用 got 发送 POST 请求代码如下:

```
1   const got = require('got');
2   exports.main = async(event, context) =>{
3     let getResponse = await got('httpbin.org/get')
4     let postResponse = await got('httpbin.org/post', {
5       method: 'POST',
6       headers: {
7         'Content-Type': 'application/json'
8       },
9       body: JSON.stringify({
10        title: 'title test',
11        value: 'value test'
12      })
13    })
14    return postResponse.body
15  }
```

3. 在微信小程序中调用云函数

在微信小程序中调用云函数,微信小程序页面 home.wxml 代码如下:

```
1   <cu-custom bgColor = "bg-gradual-pink" isBack = "{{true}}">
2     <view slot = "backText">返回</view>
3     <view slot = "content">导航栏</view>
4   </cu-custom>
5   <button bindtap = 'http'> http </button>
```

相应的 home.js 代码如下:

```
1   Page({
2     http: function (event) {
3       wx.cloud.callFunction({
4         name: 'http'
5       }).then(res =>{
6         console.log(res.result)
7       })
8     }
9   })
```

单击微信小程序 http 按钮,就可以调用云函数发送 HTTP 请求,本实例中发送 GET 请求的结果如图 5-5 所示。

发送 POST 请求结果如图 5-6 所示。

除了使用 got 依赖库可以发送 HTTP 请求外,还可以使用 request-promise 和 axios 依赖库来发送 HTTP 请求,接下来介绍如何使用 axios 来发送 HTTP 请求,request-promise 的用法与 got 和 axios 类似,这里不再详细介绍。

4. 在云函数中安装 axios 依赖库

在云函数 http 的 package.json 文件中的 dependencies 中写入需要安装的依赖库:

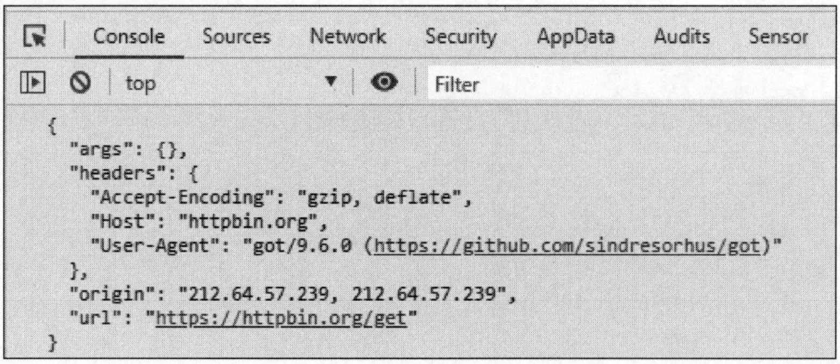

图 5-5　发送 GET 请求结果

图 5-6　发送 POST 请求结果

```
1  "dependencies": {
2    "axios": "latest"
3  }
```

5. 在云函数中使用 axios 发送 HTTP 请求

在云函数中使用 axios 发送 GET 请求代码如下：

```
1  const axios = require('axios');
2  exports.main = async (event, context) =>{
3    let getResponse = await axios.get('http://httpbin.org/get')
4    return getResponse.data
5  }
```

在云函数中使用 got 发送 POST 请求代码如下：

```
1  const axios = require('axios');
2  exports.main = async (event, context) =>{
```

```
3    let postResponse = await axios.post('http://httpbin.org/post', {
4      title: 'title test',
5      value: 'value test'
6    });
7    return postResponse.data
8  }
```

axios 和 got 发送 HTTP 请求的区别：

（1）使用 axios 必须补全路径，例如采用路径 httpbin.org/post，系统会提示错误："Error: connect ECONNREFUSED 127.0.0.1:80"；

（2）got 返回的结果为 postResponse.body，而 axios 返回的结果为 postResponse.data。

6. 云函数解析豆瓣图书信息

有较多的应用需要用到豆瓣的图书信息，但是目前豆瓣图书 API 处于停用状态，而国内很多图书 API 要么需要收费，要么就是服务器不稳定，这里使用 HTTP 请求读取豆瓣图书信息，随后对 HTML 进行解析获取图书信息，例如根据 ISBN 在豆瓣搜索的网址为 https://book.douban.com/isbn/9787302224464，网址中最后 13 位数字为图书的 ISBN 号，搜索结果如图 5-7 所示。在 Chrome 浏览器中右击，在弹出的快捷菜单中选择"查看网页源代码(V)"选项，查看网页的源代码，本节使用云函数来解析豆瓣图书信息。

图 5-7　豆瓣网根据 ISBN 搜索图书信息

因为解析豆瓣图书信息不需要使用云数据库和云存储,所以作者是直接在 Visual Studio Code(VS Code)中调试 Node.js 代码,调试界面如图 5-8 所示。

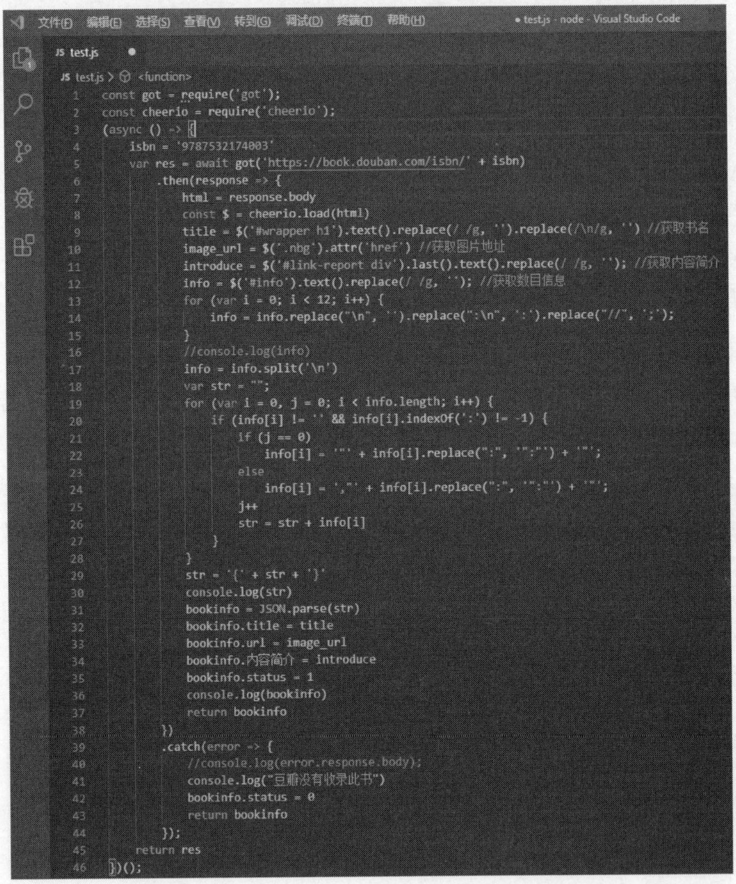

图 5-8　VS Code 调试 Node.js 代码界面

这里用到了 got 和 cheerio 依赖库,cheerio 是 Node.js 的抓取页面模块,是为服务器特别定制的,快速、灵活、实施的 jQuery 核心实现,适合各种 Web 爬虫程序。安装 got 和 cheerio 命令为"npm install got；npm install cheerio"。

在微信小程序云函数中使用 got 和 cheerio 只需要在 package.json 文件中的 dependencies 项添加依赖项:

```
1  "dependencies": {
2    "got": "^9.6.0",
3    "cheerio": "latest"
4  }
```

云函数 douban 中 index.js 代码如下:

```
1  const got = require('got')
2  const cheerio = require('cheerio');
3  exports.main = async (event, context) =>{
```

```javascript
4      var isbn = event.isbn
5      var bookinfo = {}
6      var res = await got('https://book.douban.com/isbn/' + isbn)
7        .then(response =>{
8          html = response.body
9          const $ = cheerio.load(html)
10         title = $('#wrapper h1').text().replace(/ /g,'').replace(/\n/g,'')     //获取书名
11         image_url = $('.nbg').attr('href')                                      //获取图片地址
12         introduce = $('#link-report div').last().text().replace(/ /g,'');       //获取内容简介
13         info = $('#info').text().replace(/ /g,'');                              //获取图书信息
14         for (var i = 0; i<12; i++){
15           info = info.replace("\n",'').replace(":\n",':').replace("//",';');
16         }
17         //console.log(info)
18         info = info.split('\n')
19         var str = "";
20         for (var i = 0, j = 0; i< info.length; i++) {
21           if (info[i] != '' && info[i].indexOf(':') != -1){
22             if (j == 0)
23               info[i] = '"' + info[i].replace(":", '":"') + '"';
24             else
25               info[i] = ',"' + info[i].replace(":", '":"') + '"';
26             j++
27             str = str + info[i]
28           }
29         }
30         str = '{' + str + '}'
31         console.log(str)
32         bookinfo = JSON.parse(str)
33         bookinfo.title = title
34         bookinfo.url = image_url
35         bookinfo.内容简介 = introduce
36         bookinfo.status = 1 //获取图书信息成功
37         console.log(bookinfo)
38         return bookinfo
39       })
40       .catch(error =>{
41         //console.log(error.response.body);
42         console.log("豆瓣没有收录此书")
43         bookinfo.status = 0 //获取图书信息失败
44         return bookinfo
45       });
46     return res
47   }
```

这里采用 cheerio 来解析 HTML，cheerio 的使用本书不做重点介绍，读者可阅读 cheerio 官方文档（https://cheerio.js.org/），在解析 HTML 页面时，读者应该对照查看 HTML 网页源代码。代码第 6 行向豆瓣网发送 HTTP 请求，返回页面信息；代码第 10 行解析图书书名；代码第 11 行解析图书封面图片地址；代码第 12 行解析图书内容简介；代

码第 13～31 行获取图书信息（作者、出版社、出版年、页数、定价等），因为读取的内容为字符串，需要修改成特定的格式，其中第 14～16 行去掉文本中不需要的字符；代码第 32 行把字符串转换为 JSON 格式。

在微信小程序中调用云函数，微信小程序页面 home.wxml 代码如下：

```
1  <cu-custom bgColor = "bg-gradual-pink" isBack = "{{true}}">
2      <view slot = "backText">返回</view>
3      <view slot = "content">导航栏</view>
4  </cu-custom>
5  <button bindtap = 'http'>http</button>
```

相应的 home.js 代码如下：

```
1   Page({
2     http: function (event) {
3       wx.cloud.callFunction({
4         name: 'douban',
5         data:{
6            isbn:"9787505722835"
7         }
8       }).then(res =>{
9         console.log(res.result)
10      })
11    }
12  })
```

单击微信小程序 http 按钮，就可以调用云函数解析图书对应 ISBN 号的图书信息，本实例中微信小程序调用云函数根据 ISBN 查询图书信息结果如图 5-9 所示。

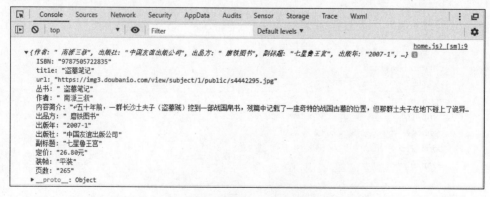

图 5-9　微信小程序根据 ISBN 查询图书信息结果

5.2　云函数将数据库数据生成 Excel

有时管理员需要把数据库中的数据导出来，方便管理员查看、核对数据，本节演示如何在云函数中使用 node-xlsx 类库把数据库数据生成 Excel。整个过程分为以下几个过程：

(1) 创建云函数,并安装 node-xlsx 类库;
(2) 读取数据库集合中的所有数据;
(3) 通过 node-xlsx 类库把数据写入 Excel;
(4) 把生成的 Excel 文件上传到云存储,供微信小程序端下载浏览。

下面分别介绍。

1. 创建云函数,并安装 node-xlsx 类库

这里采用 5.1 节中介绍的第二种方法安装依赖库,用户生成云函数后,在 package.json 文件中的 dependencies 项添加依赖项:

```
1  "dependencies": {
2    "wx-server-sdk": "latest",
3    "node-xlsx": "latest"
4  }
```

2. 读取数据库集合中的所有数据

因为数据库集合中存在较多记录,而采用微信小程序每次只能读取数据库集合中 20 条记录,云函数每次只能读取数据库集合中 100 条记录,因为有默认 100 条的限制,所以很可能一个请求无法取出所有数据,需要分批次取。微信小程序开发文档中给出了 Collection.get()取集合所有数据的方法,这里只需要把数据库集合修改成想要读取的数据库集合即可,代码如下:

```
1  let databasename = event.databasename
2  //先读取出集合记录总数
3  const countResult = await db.collection(databasename).count()
4  const total = countResult.total
5  //计算需分几次读取
6  const batchTimes = Math.ceil(total / 100)
7  //承载所有读操作的 promise 的数组
8  console.log(total, batchTimes)
9  const tasks = []
10 for (let i = 0; i < batchTimes; i++) {
11   const promise = db.collection(databasename).skip(i * MAX_LIMIT).limit(MAX_LIMIT).get()
12   tasks.push(promise)
13 }
14 let dbdata = (await Promise.all(tasks)).reduce((acc, cur) =>{
15   return {
16     data: acc.data.concat(cur.data),
17     errMsg: acc.errMsg,
18   }
19 })
```

代码第 1 行对应读取的数据库集合名称,第 2、3 行读取集合中记录总数,因为云函数读取数据库集合时每次最多只能读取 100 条记录,所以代码第 6 行计算需要分几次读取集合中的数据,代码第 9~13 行分批读取集合中的数据。代码第 14~18 行把所有分批读取的数据进行汇总。

3. 通过 node-xlsx 类库把数据写入 Excel

可以使用 node-xlsx 中的 xlsx.build([{name：excelname，data：exceldata}]) 生成 Excel 文件，其中 name 参数为需要生成的 Excel 文件名称，data 是一个二维数组，每个元素都对应一个单元格。

```
//1. 定义 Excel 文件名
let excelname = Math.floor(10000 * Math.random()) + 'test.xlsx'
//2. 定义存储的数据
let exceldata = [];
let row = ['设备ID', '设备名称', '领用人', '产商', '存放地', '价格', '购买日期', '供应商'];  //表属性
exceldata.push(row);
//console.log(dbdata.data)
for (let item of dbdata.data) {
    let arr = [];
    arr.push(item.deviceid);
    arr.push(item.devicename);
    arr.push(item.deviceuser);
    arr.push(item.manufacturer);
    arr.push(item.place);
    arr.push(item.price);
    arr.push(item.purchasedate);
    arr.push(item.supplier);
    exceldata.push(arr)
}
// console.log(exceldata)
//3. 把数据保存到 Excel 文件中
var buffer = await xlsx.build([{
    name: excelname,
    data: exceldata
}]);
```

代码第 2 行采用随机数＋test.xls 方式生成 Excel 文件名称，防止多次生成 Excel 文件在存储中出现命名冲突。这里采用逐行数据写入 exceldata，其中第 5 行写入 Excel 第一行数据的 title；代码第 8～19 行把每条记录转换成 Excel 中的一行数据；代码第 22～25 行生成 Excel 文件。

4. 把生成的 Excel 文件上传到云存储

最后为了提供微信小程序用户下载，需要把生成的 Excel 文件上传到云存储，代码如下：

```
await cloud.uploadFile({
    cloudPath: excelname,
    fileContent: buffer,    //excel 二进制文件
})
```

完整的数据库集合生成 Excel 文件的云函数代码如下：

```javascript
const cloud = require('wx-server-sdk')
cloud.init()
const db = cloud.database()
var xlsx = require('node-xlsx');
const MAX_LIMIT = 100
exports.main = async (event, context) =>{
  let databasename = event.databasename
  //先读取出集合记录总数
  const countResult = await db.collection(databasename).count()
  const total = countResult.total
  //计算需分几次读取
  const batchTimes = Math.ceil(total / 100)
  //承载所有读操作的 promise 的数组
  console.log(total, batchTimes)
  const tasks = []
  for (let i = 0; i< batchTimes; i++) {
    const promise = db.collection(databasename).skip(i * MAX_LIMIT).limit(MAX_LIMIT).get()
    tasks.push(promise)
  }
  let dbdata = (await Promise.all(tasks)).reduce((acc, cur) =>{
    return {
      data: acc.data.concat(cur.data),
      errMsg: acc.errMsg,
    }
  })

  //1. 定义 Excel 文件名
  let excelname = Math.floor(10000 * Math.random()) + 'test.xlsx'
  //2. 定义存储的数据
  let exceldata = [];
  let row = ['设备ID','设备名称','领用人','产商','存放地','价格','购买日期','供应商'];//表头
  exceldata.push(row);
  //console.log(dbdata.data)
  for (let item of dbdata.data) {
    let arr = [];
    arr.push(item.deviceid);
    arr.push(item.devicename);
    arr.push(item.deviceuser);
    arr.push(item.manufacturer);
    arr.push(item.place);
    arr.push(item.price);
    arr.push(item.purchasedate);
    arr.push(item.supplier);
    exceldata.push(arr)
  }
  // console.log(exceldata)
  //3. 把数据保存到 Excel 文件中
  var buffer = await xlsx.build([{
```

```
49        name: excelname,
50        data: exceldata
51    }]);
52    //4. 把 Excel 文件保存到云存储中
53    return await cloud.uploadFile({
54        cloudPath: excelname,
55        fileContent: buffer,      //Excel 二进制文件
56    })
57 }
```

这里需要说明的是，调用云函数把数据库集合数据生成 Excel 文件之前需要先建立相应的数据库集合并插入数据记录，为了便于演示，这里新建了 device 数据库集合，数据从 device.json 文件中导入 104 条设备记录。调用云函数成功后，打开云开发控制台，选择"存储"→"存储管理"选项，可以看到新生成的 Excel 文件，如图 5-10 所示。

图 5-10　云函数生成的 Excel 文件

最后在微信小程序端调用云函数就可以下载并浏览 Excel 文件了，可以在微信小程序端加一个按钮，按钮的 bindtap 事件 generateExcel 代码如下：

```
1  generateExcel: function (event){
2    wx.cloud.callFunction({
3      name: 'excel',
4      data:{
5        databasename:"device"
6      }
7    }).then(res =>{
8      console.log(res.result.fileID)
9      wx.showLoading({
10        title: '文档打开中...',
11      })
12      wx.cloud.downloadFile({
13        fileID: res.result.fileID
14      }).then(res =>{
15        // get temp file path
16        console.log(res.tempFilePath)
17        const filePath = res.tempFilePath
18        wx.openDocument({
19          filePath: filePath,
```

```
20            success: res =>{
21                console.log('打开文档成功')
22                wx.hideLoading()
23            }
24        })
25    }).catch(error =>{
26        // handle error
27    })
28  })
29 }
```

代码第 2~7 行调用云函数,云函数名称为 excel,传递参数数据库集合名称 device。调用云函数生成 Excel 文件后,通过 wx.cloud.downloadFile()从云存储中下载 Excel 文档,随后使用 wx.openDocument()打开下载的文档进行查看。

有些读者可能会尝试通过云函数把 Excel 数据导入到云数据库中,目前微信小程序端和云函数端 API 接口还不支持数据库数据导入功能(通过 cloudbase-manager-node 依赖库目前支持数据库导入功能,不过仅限于导入 CSV 和 JSON 格式的文件,详情见 https://github.com/TencentCloudBase/cloudbase-manager-node)。Collection.add()一次操作只能插入一条记录,不支持批量插入数据,因此插入记录的方式为:"打开数据库"→"插入一条记录"→"关闭数据库"→"打开数据库"→"插入一条记录"→"关闭数据库",如此往复,当数据量比较大时,逐条记录插入数据库操作的资源消耗是比较大的,云函数会提示错误:{"errCode":-501004,"errMsg": "\[LimitExceeded.NoValidConnection] Connection num overrun."}。当数据量比较大时,通过云函数从 Excel 中读取数据,然后用 Collection.add()逐条插入数据库的方法并不可行,而且默认情况下云函数超时时间为 3s(云开发控制台上可以设置云函数超时时间最长为 20s),也就是说云函数在 3s 内没有返回结果就会超时;如果读者一定要从 Excel 中导入数据库,目前可以通过 HTTP API 数据库导入接口导入数据,见 15.2.4 节通过 HTTP API 接口批量插入数据。

5.3 本地调试

云开发提供了云函数本地调试功能,在本地提供了一套与线上一致的 Node.js 云函数运行环境,让开发者可以在本地对云函数调试,使用本地调试可以提高开发、调试效率。

- 单步调试/断点调试:比起通过云开发控制台中查看线上打印的日志的方法进行调试,使用本地调试后可以对云函数 Node.js 实例进行单步调试/断点调试。
- 集成微信小程序测试:在模拟器中对微信小程序发起的交互点击等操作如果触发了开启本地调试的云函数,会请求到本地实例而不是云端。
- 优化开发流程,提高开发效率:调试阶段不需上传部署云函数,在调试云函数时,相对于不使用本地调试时的调试流程("本地修改代码"→"上传部署云函数"→"调用")的调试流程,省去了上传等待的步骤,改成只需"本地修改"→"调用"的流程,大大提高开发、调试效率。

同时,本地调试还定制化提供了特殊的调试能力,包括 Network 面板支持展示 HTTP

请求和云开发请求、调用关系图展示、本地代码修改时热重载等能力,帮助开发者更好地开发、调试云函数。建议开发者在开发阶段和上传代码前先使用本地调试测试通过后再上线部署。

在 cloudfunctions 文件上右击,在弹出的快捷菜单中选择"本地调试"选项,开发者可通过右击云函数名唤起本地调试界面。在本地调试界面中点击相应云函数并勾选"开启本地调试"复选框方可进行该云函数的本地调试,如图 5-11 所示。取消勾选"开启本地调试"复选框后可关闭对该云函数的本地调试。若云函数中使用到 npm 模块,需在云函数本地目录安装相应依赖才可正常使用云函数本地调试功能。在开启本地调试的过程中,系统会检测该云函数本地是否已安装了 package.json 中所指定的依赖库,如果没有安装相应的依赖库则会给出错误,出现错误提示:"Error:Cannot find module 'wx-server-sdk'…"。因为在进行云函数开发的时候,首先就会引用 const cloud = require('wx-server-sdk')。云环境会自动安装 package.json 中所指定的依赖库,问题是当前项目本地调试环境还没有这个模块。所以,需要先安装这个模块。解决方法:在本地安装 wx-server-sdk 依赖,如图 5-12 所示,在 cloudfunctions 文件上右击,在弹出的快捷菜单中选择"在终端中打开"选项,输入命令 npm install-save wx-server-sdk@latest,安装好依赖库以后进入云函数本地调试窗口,勾选"开启本地调试"复选框。如果用户要使用本地调试,在安装依赖库时建议使用方法一通过 npm install 安装依赖库,这样本地目录中会安装依赖库。

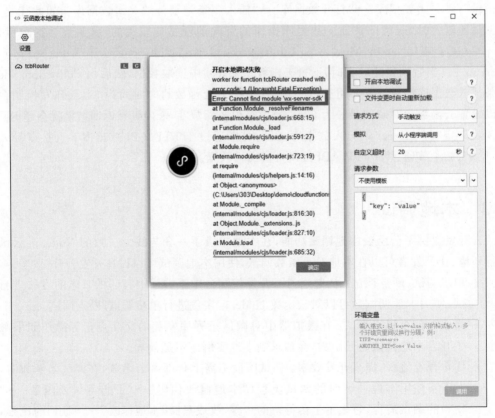

图 5-11 开启本地调试错误

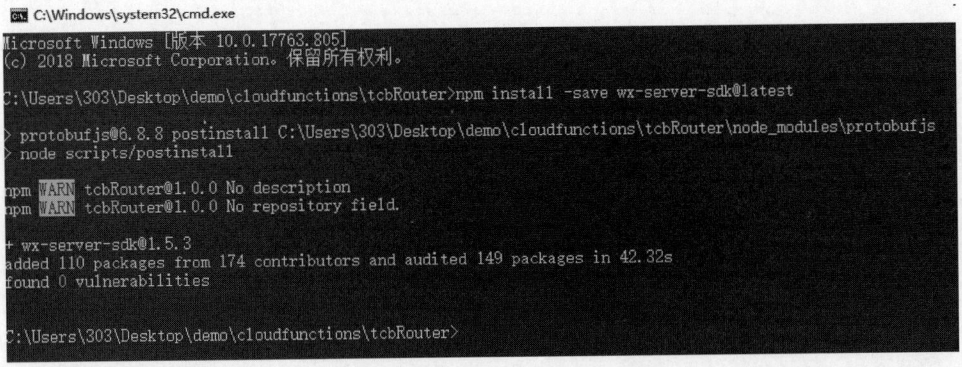

图 5-12 安装 wx-server-sdk 依赖

对于已开启本地调试的云函数，微信开发者工具模拟器中对该云函数的请求以及其他开启了本地调试的云函数对该云函数的请求，都会自动请求到该本地云函数实例。为方便调试，一个云函数在本地仅会有一个实例，实例会串行处理请求，本地云函数递归调用自身将被拒绝。本地调试云函数实例右侧的面板中可以开启"文件变更时自动重新加载"，开启后，每当函数代码发生修改，就会自动重新加载云函数实例，这就省去了关闭本地调试再重新打开本地调试开关的麻烦。

本地调试使用须知：

- npm 依赖：若云函数中使用到 npm 模块，需在云函数本地目录安装相应依赖才可正常使用云函数本地调试功能。
- native 依赖：如果云函数中使用了 native 依赖（注意 native 依赖是需要各个平台分别编译的），比如在 Windows 上本地调试时安装的 native 依赖是 Windows 上编译的结果，而线上云函数环境是 Linux 环境，因此调试完毕上传云函数注意选择云端安装依赖的上传方式，该方式会自动在云端环境下编译 native 依赖，如果由于云端编译环境不足而需要选择全量上传则需要在 Linux CentOS 7 下编译后上传结果。
- Node.js 版本：系统默认使用开发者工具自带的 Node.js，用户可通过点击本地调试面板左上方的设置进行修改。
- 云函数实例个数：本地调试下一个云函数最多只会有一个实例，对本地云函数实例的并发请求会被实例串行处理。

5.4 定时触发器

如果云函数需要定时/定期执行，也就是定时触发，可以使用云函数定时触发器。配置了定时触发器的云函数，会在相应时间点被自动触发，函数的返回结果不会返回给调用方。

在需要添加触发器的云函数目录下新建文件 config.json，格式如下：

```
1  {
2    // triggers 字段是触发器数组,目前仅支持一个触发器,即数组只能填写一个,不可添加多个
3    "triggers": [
4      {
5        // name：触发器的名字,规则见下方说明
6        "name": "myTrigger",
7        // type：触发器类型,目前仅支持 timer（即定时触发器）
8        "type": "timer",
9        // config：触发器配置,在定时触发器下,config 格式为 Cron 表达式,规则见下方说明
10       "config": "0 0 2 1 * * *"
11     }
12   ]
13 }
```

字段规则如下：

- 定时触发器名称(name)：最大支持 60 个字符,支持 a～z、A～Z、0～9、-和_。必须以字母开头,且一个函数下不支持同名的多个定时触发器。
- 定时触发器触发周期(config)：指定的函数触发时间。填写自定义标准的 Cron 表达式来决定何时触发函数。

1. Cron 表达式

Cron 表达式有 7 个必需字段,按空格分隔,如表 5-1 所示。

表 5-1 Cron 表达式的必需字段

第1个	第2个	第3个	第4个	第5个	第6个	第7个
秒	分钟	小时	日	月	星期	年

其中,每个字段都有相应的取值范围,如表 5-2 所示。

表 5-2 必需字段相应的取值范围

字段	值	通配符
秒	0～59 的整数	, - * /
分钟	0～59 的整数	, - * /
小时	0～23 的整数	, - * /
日	1～31 的整数(需要考虑月的天数)	, - * /
月	1～12 的整数 或 JAN、FEB、MAR、APR、MAY、JUN、JUL、AUG、SEP、OCT、NOV、DEC	, - * /
星期	0～6 的整数或 MON、TUE、WED、THU、FRI、SAT、SUN。其中,0 指星期一,1 指星期二,以此类推	, - * /
年	1970～2099 的整数	, — * /

2. 通配符(见表 5-3)

表 5-3 通配符

通配符	含 义
,(逗号)	代表用逗号隔开的字符的并集。例如,在"小时"字段中 1,2,3 表示 1 点、2 点和 3 点

续表

通配符	含 义
—(破折号)	包含指定范围的所有值。例如,在"日"字段中,1~15 包含指定月份的 1~15 号
*(星号)	表示所有值。在"小时"字段中,* 表示每小时
/(正斜杠)	指定增量。在"分钟"字段中,输入 1/10 以指定从第一分钟开始的每隔 10 分钟重复。例如,第 11 分钟、第 21 分钟和第 31 分钟,以此类推

3. 注意事项

在 Cron 表达式中的"日"和"星期"字段同时指定值时,两者为"或"关系,即两者的条件分别均生效。

4. 示例

下面展示了一些 Cron 表达式和相关含义的示例:

- */5 * * * * * 表示每 5 秒触发一次;
- 0 0 2 1 * * * 表示在每月的 1 日的凌晨 2 点触发;
- 0 15 10 * * MON-FRI * 表示在周一到周五每天上午 10:15 触发;
- 0 0 10,14,16 * * * 表示在每天上午 10 点、下午 2 点和 4 点触发;
- 0 */30 9-17 * * * 表示在每天上午 9 点到下午 5 点内每半小时触发;
- 0 0 12 * * WED * 表示在每个星期三中午 12 点触发。

5.5 云函数高级用法——TcbRouter

微信小程序云开发的云函数都是运行在不同的开发环境中,每个云函数都是一个功能模块,传统的云函数用法是一个云函数处理一个任务,如图 5-13 所示。但是一个用户在一个环境中只有 50 个云函数,经常会出现 50 个云函数不够用的情况,而且为了方便维护管理和公用代码块复用,需要将具有相似的处理逻辑云函数合并成一个云函数,这时就需要用到路由控制。一种常见的方法是在请求云函数时多加一个参数,在云函数中根据该参数利用 switch…case 语句(或者 if…else 语句)来区别对该云函数的不同请求,这种方法比较简单粗暴,例如:

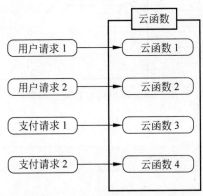

图 5-13 一个云函数处理一个任务

```
1   exports.main = async(event, context) =>{
2       let action = event.action
3       switch (action) {
4         case 'actionA':
5           {
6               //执行 actionA 任务
7           }
8         case 'actionB':
9           {
10              //执行 actionB 任务
11          }
12        case 'actionC':
13          {
14              //执行 actionC 任务
15          }
16        default:
17          {
18              //执行 default 任务
19          }
20      }
21  }
```

在上面的例子中，请求云函数时多加一个参数 action，就可以在云函数中接受参数 action，然后使用 switch…case 语句来执行不同的任务。

switch…case 的处理方式往往可读性差，不利于管理。为了解决这个问题，腾讯云 Tencent Cloud Base 团队开发了 TcbRouter，TcbRouter 是一个基于 Koa 风格的云函数路由库，通过 TcbRouter 路由管理云函数可以优化云函数处理逻辑，如图 5-14 所示。云函数中有一个分派任务的路由管理，将不同的任务分配给不同的本地函数处理。

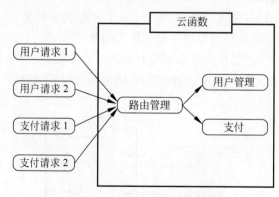

图 5-14　通过路由管理云函数

TcbRouter 安装命令如下：

```
1   npm install -- save tcb-router
```

TcbRouter 框架如下：

```
1  const TcbRouter = require('tcb-router');
2  //云函数入口函数
3  exports.main = async (event, context) =>{
4    const app = new TcbRouter({ event });
5    //-----------------------------------------
6    //使用app.use()或者app.router()处理路由
7    //-----------------------------------------
8    return app.serve();
9  }
```

app.use()可以在所有的路由上进行处理：

```
1  //app.use()表示该中间件,适用于所有的路由
2  app.use(async (ctx, next) =>{
3      ctx.data = {};
4      await next();      //执行下一中间件
5  });
```

app.use()方法是支持异步的，所以为了保证正常的按照洋葱模型的执行顺序执行代码，需要在调用 next() 时让代码等待，等待异步结束后再继续向下执行，所以在使用 TcbRouter 时建议使用 async/await。

app.router() 可以处理某个特定的路由，也可以处理路由为数组的情况。路由为数组的情况如下：

```
1  //路由为数组表示,该中间件适用于 user 和 timer 两个路由
2  app.router(['user', 'timer'], async (ctx, next) =>{
3      ctx.data.company = 'Tencent';
4      await next();      //执行下一中间件
5  });
```

路由为字符串，适用于处理某个特定的路由：

```
1   app.router('user', async (ctx, next) =>{
2       ctx.data.name = 'heyli';
3       await next(); //执行下一中间件
4   }, async (ctx, next) =>{
5       ctx.data.sex = 'male';
6       await next(); //执行下一中间件
7   }, async (ctx) =>{
8       ctx.data.city = 'Foshan';
9       // ctx.body 返回数据到微信小程序端
10      ctx.body = { code: 0, data: ctx.data};
11  });
```

微信小程序端调用：

```
1    //调用名为 router 的云函数,路由名为 user
2    wx.cloud.callFunction({
3        //要调用的云函数名称
4        name: "router",
5        //传递给云函数的参数
6        data: {
7            $url: "user", //要调用的路由的路径,传入准确路径或者通配符 *
8            other: "xxx"
9        }
10   });
```

TcbRouter 演示案例:

在主页(pages/home/home)中,home.wxml 页面里放两个按钮,分别添加两个 bindtap 事件,为 school 和 user:

```
1    <cu-custom bgColor="bg-gradual-pink" isBack="{{false}}">
2        <view slot="backText">返回</view>
3        <view slot="content">TcbRouter 演示</view>
4    </cu-custom>
5
6    <button type="primary" bindtap="school">school</button>
7    <button type="primary" bindtap="user">user</button>
```

TcbRouter 演示页面如图 5-15 所示。相应的 home.js 代码如下:

图 5-15　TcbRouter 演示页面

```
1    Page({
2        school: function (event) {
3            //调用名为 tcbRouter 的云函数,路由名为 school
```

```
4      wx.cloud.callFunction({
5        //要调用的云函数名称
6        name: "tcbRouter",
7        //传递给云函数的参数
8        data: {
9          $url: "school",              //要调用的路由的路径,传入准确路径或者通配符 *
10         other: "xxx"
11       }
12     }).then(res =>{
13       console.log(res)
14     });
15   },
16   user: function (event) {
17                                      //调用名为 tcbRouter 的云函数,路由名为 user
18     wx.cloud.callFunction({
19                                      //要调用的云函数名称
20       name: "tcbRouter",
21                                      //传递给云函数的参数
22       data: {
23         $url: "user",                //要调用的路由的路径,传入准确路径或者通配符 *
24         other: "xxx"
25       }
26     }).then(res =>{
27       console.log(res)
28     });
29   }
30 })
```

代码第 2~15 行的 school 事件中,调用 tcbRouter 云函数,路由路径为 school;代码第 16~29 行的 user 事件中,调用 tcbRouter 云函数,路由路径为 user。

在 cloudfunctions 上右击,在弹出的快捷菜单中选择"新建 Node.js 云函数"选项并命名为 tcbRouter,然后右击,在弹出的快捷菜单中选择"在终端中打开"选项,如图 5-16 所示。在终端输入 npm install --save tcb-router,终端安装 tcbRouter 完成后的效果如图 5-17 所示。

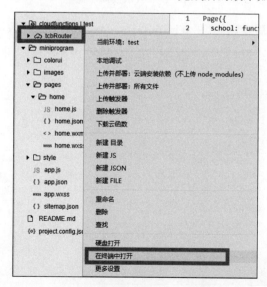

图 5-16　新建 tcbRouter 云函数

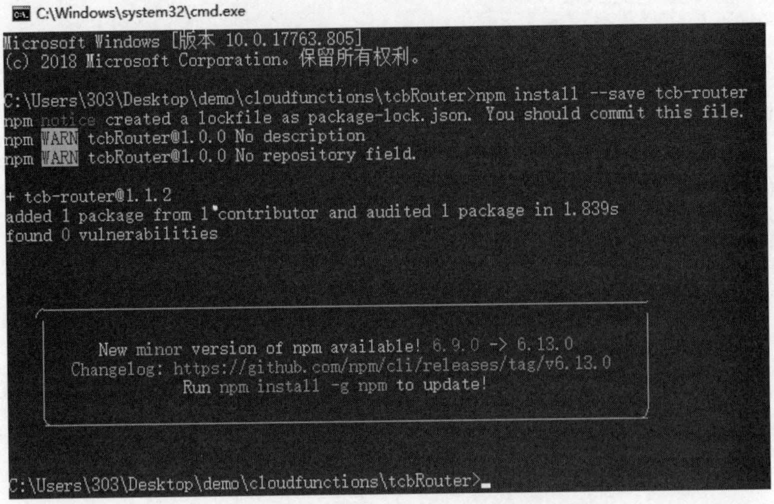

图 5-17　终端安装 tcbRouter 完成后的效果

安装完 tcbRouter 后，在云函数 tcbRouter/package.json 中可以看到该云函数的依赖包，如图 5-18 所示，如果 tcbRouter 安装成功，在这里就会显示安装的版本。

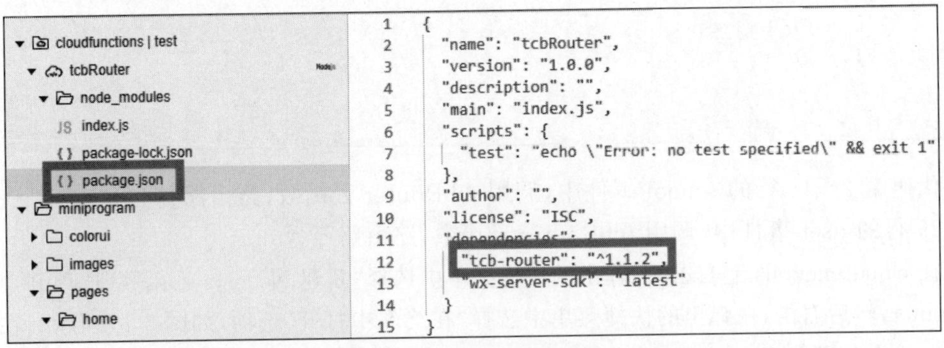

图 5-18　云函数依赖包

安装完 tcbRouter 后，就可以在 tcbRouter/index.js 中写云函数了，云函数的代码如下：

```
1   //云函数入口文件
2   const cloud = require('wx-server-sdk')
3   const TcbRouter = require('tcb-router')
4   cloud.init()
5   
6   //云函数入口函数
7   exports.main = async (event, context) =>{
8     const app = new TcbRouter({ event })
9   
10    // app.use()表示该中间件会适用于所有的路由
11    app.use(async (ctx, next) =>{
12      console.log('---------->进入全局的中间件')
13      ctx.data = {};
14      ctx.data.openId = event.userInfo.openId
```

```
15        await next();                              //执行下一中间件
16        console.log('---------->退出全局的中间件')
17    });
18
19    //路由为数组表示,该中间件适用于user和school两个路由
20    app.router(['user', 'school'], async (ctx, next) =>{
21        console.log('---------->进入数组路由中间件')
22        ctx.data.from = '微信小程序云函数实战'
23        await next();                              //执行下一中间件
24        console.log('---------->退出数组路由中间件')
25    });
26
27    //路由为字符串,该中间件只适用于user路由
28    app.router('user', async (ctx, next) =>{
29        console.log('---------->进入用户路由中间件')
30        ctx.data.name = 'xiaoqiang user';
31        ctx.data.role = 'Developer'
32        await next();
33        console.log('---------->退出用户路由中间件')
34    }, async (ctx) =>{
35        console.log('---------->进入用户昵称路由中间件')
36        ctx.data.nickName = 'BestTony'
37        ctx.body = { code: 0, data: ctx.data };   //将数据返回云函数,用ctx.body
38        console.log('---------->退出用户昵称路由中间件')
39    });
40
41    app.router('school', async (ctx, next) =>{
42        ctx.data.name = '腾讯云学院';
43        ctx.data.url = 'cloud.tencent.com'
44        await next();
45    }, async (ctx) =>{
46        ctx.data.nickName = '学院君'
47        ctx.body = { code: 0, data: ctx.data };   //将数据返回云函数,用ctx.body
48    });
49
50    return app.serve();
51 }
```

在云函数tcbRouter上右击,在弹出的快捷菜单中选择"上传并部署:上传安装依赖(不上传node_modules),等待上传云函数结束"选项。

单击图5-15中user按钮,在微信开发者工具中查看Console窗口输出结果,微信小程序端tcbRouter输出结果如图5-19所示。从result.data数据中可以看出,单击user按钮,在云函数tcbRouter中依次进入了"app.use中间件"→"app.router(['user', 'school'])路由"→"app.router('user')路由"。

tcbRouter每个中间件默认接收两个参数:第一个参数是Context对象;第二个参数是next函数。只要调用next函数,就可以把执行权转交给下一个中间件。tcbRouter的精粹思想

```
▼{errMsg: "cloud.callFunction:ok", result: {…}, requestID: "9880e5e2-07a0-11ea-9b52-525400b2c41b"}
   errMsg: "cloud.callFunction:ok"
   requestID: "9880e5e2-07a0-11ea-9b52-525400b2c41b"
  ▼result:
     code: 0
    ▶ data: {openId: "okzLz5Eqw60N0Mt0eSDx3dhs1fyk", from: "小程序云函数实战", name: "xiaoqiang user", role: "Developer", nickName: "BestTony"}
    ▶ __proto__: Object
  ▶ __proto__: Object
```

图 5-19 微信小程序端 tcbRouter 输出结果

就是洋葱模型（中间件模型），如图 5-20 所示。多个中间件会形成一个栈结构（middle stack），以"先进后出"（first-in-last-out）的顺序执行。整个过程就像先入栈后出栈的操作。

图 5-20 tcbRouter 洋葱模型

为了更好地理解 tcbRouter 洋葱模型，开发者可以进入云开发控制台，选择"云函数"→"日志"→tcbRouter 选项，可以看到云函数 tcbRouter 中间件的执行顺序，如图 5-21 所示。

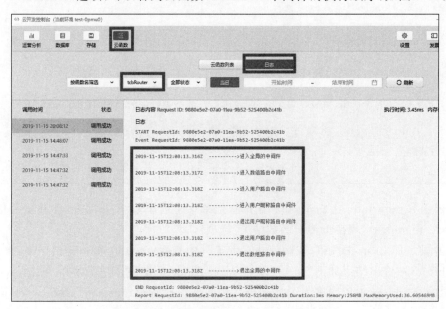

图 5-21 云函数 tcbRouter 中间件执行顺序

tcbRouter 中间件模型非常好用并且简洁，但是也有自身的缺陷，一旦中间件数组过于庞大，性能会有所下降，因此在实际项目开发中，并不是把所有请求都放在同一个云函数中。

第二部分

微信小程序云开发实战

第 6 章

新闻微信小程序

新闻微信小程序包含如下 4 个页面：

```
1  "pages": [
2    "pages/home/home",              //新闻主页
3    "pages/addnews/addnews",        //添加新闻页面
4    "pages/newsdetail/newsdetail",  //新闻详情页
5    "pages/auth/auth"               //授权页面
6  ],
```

新闻微信小程序还包含 2 个数据库集合 news 和 newscomment，权限均为"所有用户可读，仅创建者可读写"。

6.1 授权页面

之前微信授权登录时是直接通过 getUserInfo 接口弹出授权窗口。由于微信官方修改了 getUserInfo 接口，所以现在无法实现一进入微信小程序就弹出授权窗口，只能使用 button 组件，并将 open-type 指定为 getUserInfo 类型，获取用户基本信息。从 2019 年 9 月 1 日开始，微信小程序平台对微信小程序内的账号登录功能进行规范，在用户清楚知悉、了解微信小程序的功能之前，要求用户进行账号登录，包括但不限于打开微信小程序后立即跳转提示登录或打开微信小程序后立即强制弹窗要求登录，都属于违反要求的情况；任何微信小程序调用账号登录功能，应当为用户清晰提供可取消或拒绝的选项按钮，不得以任何方式强制用户进行账号登录。也就是说任何一个微信小程序页面不能一打开就跳转到登录页面，而且登录页面要有允许用户取消或拒绝的选项。

本案例演示通过 getUserInfo 获得用户信息授权页面（pages/auth/auth），样式 auth.wxml 如下：

```
1   <cu-custom bgColor="bg-gradual-pink" isBack="{{false}}">
2     <view slot="content">登录</view>
3   </cu-custom>
4
5   <view class="bg-white solid-bottom">
6     <view class="title">
7       <open-data type="userAvatarUrl"></open-data>
8     </view>
9   </view>
10
11  <view class="padding bg-white">
12    <view class="text-black">该应用将获取以下授权:</view>
13    <view class="text-grey">获取您的公开信息(昵称,头像,性别等)</view>
14  </view>
15
16  <view style='display: flex; justify-content: center; bg-white'>
17  <button bindtap="refuseAuth" class="cu-btn xl bg-green padding-sm margin-xs">拒绝</button>
18  <button bindgetuserinfo="bindGetUserInfo" open-type='getUserInfo' class="cu-btn xl
19  bg-green padding-sm margin-xs">允许</button>
20  </view>
```

代码第11～14行提示用户将授权微信昵称、头像、性别等信息。代码第16～20行,让用户选择拒绝授权或者接受授权。需要说明的是,允许授权button的open-type必须指定为getUserInfo类型,bindgetuserinfo事件会返回获取到的用户信息。授权页面的显示效果如图6-1所示。

图6-1 新闻微信小程序授权页面的显示效果

相应的 auth.js 代码如下：

```
1   Page({
2     refuseAuth: function () {
3       wx.switchTab({
4         url: '../home/home'
5       })
6     },
7     bindGetUserInfo: function (e) {
8       if (e.detail.userInfo) {
9         wx.setStorageSync('userInfo', e.detail.userInfo);
10        wx.showToast({
11          title: '授权成功...',
12          icon: 'none',
13          duration: 1500
14        })
15        wx.navigateBack({
16          delta: 1
17        })
18      }
19    },
20  })
```

代码第 2~6 行，当用户单击"拒绝"按钮时，页面跳转到微信小程序首页（pages/home/home）；代码第 7~19 行，当用户授权成功后，把用户信息存入本地缓存，供其他页面调用，并显示授权成功消息提示框，退回上一页面，用户授权登录后 Storage 缓存的数据如图 6-2 所示。

```
- Object
    nickName: "ii"
    gender: 1
    language: "zh_CN"
    city: "Wenzhou"
    province: "Zhejiang"
    country: "China"
    avatarUrl: "https://wx.qlogo.cn/mmopen/vi_32/DYAIOgq83epTozkia3GuZliaWPT9yMPsNcRAVEjzYa5W4NV5ShAaUbVEQvc1XVO7pm... <string is too large>"
```

图 6-2　授权登录后 Storage 缓存的数据

6.2　添加新闻页面

添加新闻页面（pages/addnews/addnews）上传图片样式如图 6-3 所示，采用 ColorUI 中的表单样式（代码见 pages/component/form），修改后的 addnews.wxml 代码如下：

```
1  <cu-custom bgColor = "bg-gradual-pink" isBack = "{{false}}">
2    <view slot = "backText">返回</view>
3    <view slot = "content">添加新闻</view>
4  </cu-custom>
5
6  <form>
```

```
7      <view class="cu-form-group">
8        <view class="title">
9          <text style="color:red">*</text>新闻标题:</view>
10       <input bindinput="updateValue" data-name='title' placeholder="请输入新闻标题"></input>
11     </view>
12     <view class="cu-form-group align-start">
13       <view class="title">
14         <text style="color:red">*</text>新闻内容:</view>
15       <textarea maxlength="-1" bindinput="updateValue" data-name='content' placeholder=
16 "输入新闻内容"></textarea>
17     </view>
18     <view class="cu-form-group">
19       <view class="title">
20         是否置顶:</view>
21       <switch class="orange radius sm" bindchange="switchChange"></switch>
22     </view>
23
24     <view class="cu-bar bg-white margin-top">
25       <view class="action">
26         图片上传
27       </view>
28       <view class="action">
29         {{imgList.length}}/3
30       </view>
31     </view>
32     <view class="cu-form-group">
33       <view class="grid col-3 grid-square flex-sub">
34         <view class="bg-img" wx:for="{{imgList}}" wx:key="{{index}}" bindtap="ViewImage"
35 data-url="{{imgList[index]}}">
36           <image src='{{imgList[index]}}' mode='aspectFill'></image>
37           <view class="cu-tag bg-red" catchtap="DelImg" data-id="{{index}}">
38             <text class="cuIcon-close"></text>
39           </view>
40         </view>
41         <view class="solids" bindtap="ChooseImage" wx:if="{{imgList.length<3}}">
42           <text class="cuIcon-cameraadd"></text>
43         </view>
44       </view>
45     </view>
46 </form>
47 <view class="flex padding justify-center">
48     <button class="cu-btn lg bg-pink" bindtap="submitform">提交</button>
49 </view>
```

代码第 7~22 行对应表单中的输入框;代码第 24~45 行为图片上传控件(采用 ColorUI 表单中的样式),在本案例中允许用户最多上传 3 张图片。添加新闻页面如图 6-3 所示,用户可以在上传的图片右上角点击"×"按钮,对上传的图片进行删除,也可以点击图片进行全屏浏览。

图 6-3 添加新闻页面

相应的 addnews.js 代码如下：

```
1   var util = require('../../utils/util.js')
2   const db = wx.cloud.database()
3   Page({
4     data: {
5       imgList: [],
6       newsInfo: {istop: 0},
7       userInfo: {}
8     },
9
10    onShow:function()
11    {
12      var userInfo = wx.getStorageSync('userInfo');
13      if (userInfo == "") {
14        wx.navigateTo({
15          url: '../auth/auth'
16        })
17      } else {
18        this.setData({
19          userInfo: userInfo
20        })
21      }
22    },
23
24    switchChange: function(e) {
```

```js
25      let newsInfo = this.data.newsInfo
26      if (e.detail.value == true) {
27        newsInfo.istop = 1
28      } else {
29        newsInfo.istop = 0
30      }
31      this.setData({
32        newsInfo: newsInfo
33      })
34    },
35
36    updateValue: function(event) {
37      let name = event.currentTarget.dataset.name;
38      let newsInfo = this.data.newsInfo
39      newsInfo[name] = event.detail.value
40      this.setData({
41        newsInfo: newsInfo
42      })
43    },
44
45    ChooseImage() {
46      wx.chooseImage({
47        count: 3,                                      //默认为9
48        sizeType: ['original', 'compressed'],          //可以指定是原图还是压缩图,默认二者都有
49        sourceType: ['album'],                         //从相册选择
50        success: (res) =>{
51          wx.showLoading({
52            title: '图片上传中...',
53          })
54          var time = util.formatTimestring(new Date());
55          var FilePaths = []
56          const temFilePaths = res.tempFilePaths
57          let promiseArr = [];
58          for (let i = 0; i< temFilePaths.length; i++) {
59            let promise = new Promise((resolve, reject) =>{
60              var randstring = Math.floor(Math.random() * 1000000).toString() + '.png'
61              randstring = time + '-' + randstring
62              wx.cloud.uploadFile({
63                cloudPath: 'newsImages/' + randstring,
64                filePath: temFilePaths[i],     // 文件路径
65                success: res =>{
66                  // get resource ID
67                  console.log(res.fileID)
68                  FilePaths[i] = res.fileID
69                  resolve(res);
70                },
71                fail: err =>{
72                  reject(error);
73                }
```

```
 74              })
 75            })
 76            promiseArr.push(promise)
 77          }
 78
 79          Promise.all(promiseArr).then((result) =>{
 80            if(this.data.imgList.length != 0) {
 81              this.setData({
 82                imgList: this.data.imgList.concat(FilePaths)
 83              })
 84            } else {
 85              this.setData({
 86                imgList: FilePaths
 87              })
 88            }
 89            wx.hideLoading()
 90          })
 91        }
 92      });
 93    },
 94
 95    ViewImage(e) {
 96      wx.previewImage({
 97        urls: this.data.imgList,
 98        current: e.currentTarget.dataset.url
 99      });
100    },
101
102    DelImg: function(event) {
103      console.log(event.currentTarget.dataset.id)
104      var id = event.currentTarget.dataset.id
105      var imgList = this.data.imgList
106      wx.cloud.deleteFile({
107        fileList: [imgList[id]]
108      }).then(res =>{
109        imgList.splice(id, 1)
110        this.setData({
111          imgList: imgList
112        })
113        console.log(res.fileList)
114      }).catch(error =>{
115        // handle error
116      })
117    },
118
119    submitform: function(event) {
120      let newsInfo = this.data.newsInfo
121      let userInfo = this.data.userInfo
122      db.collection('news').add({
```

```
123          // data 字段表示需新增的 JSON 数据
124          data: {
125              title: newsInfo.title,
126              content: newsInfo.content,
127              publisher: userInfo.nickName,
128              imgUrl: this.data.imgList,
129              avatarUrl: userInfo.avatarUrl,
130              submitdate: db.serverDate(),
131              istop: userInfo.istop,
132              comment: 0,
133              appreciate: 0,
134              view: 0
135          }
136      })
137      .then(res =>{
138          wx.reLaunch({
139              url: '../home/home'
140          })
141      })
142    },
143  })
```

　　微信小程序 Tab 切换后,不会执行 onLoad 操作,Tab 切换对应的生命周期如表 2-5 所示。而添加新闻页面需要检测用户是否授权登录,因此需要在 onShow 事件中检测用户是否授权登录,读者要理解 onLoad(页面创建时执行)和 onShow(页面出现在前台时执行)事件的区别。代码第 10～22 行在 onShow 事件中,读取本地缓存中是否有 userInfo 信息,如果没有则跳转到授权页面。代码第 24～34 行读取开关选择器 switch 的值。代码第 36～43 行读取输入框的值。代码第 45～93 行对应用户上传图片,chooseImage()从相册中选取少于 3 张图片,为了防止图片冲突,上传的图片以日期＋6 位随机数作为文件名,通过 wx.cloud.uploadFile()上传图片,由于 wx.cloud.uploadFile()操作是异步的,在这种情况下图片上传完成后输出的顺序、判断图片是否全部上传完成等,都是一些需要注意的点。在 Promise.all()将多个 Promise 实例包装成一个新的 Promise 实例(简单来说就是当所有的异步请求成功后才会执行),在这个方法的帮助下,这些问题都迎刃而解。代码第 79～90 行通过 Promise.all()的方法,把用户上传的多张图片放到 imgList 中。代码第 95～100 行可以全屏查看用户上传的图片。代码第 102～117 行,用户单击图片右上角的"×"按钮,使用 wx.cloud.deleteFile()对云存储上的照片进行删除。代码第 119～142 行把用户填写的信息以及上传图片的地址写入云数据库 news 集合中。填写添加新闻页面后 AppData 数据如图 6-4 所示。

　　添加新闻成功后,可以在云开发控制台上进入存储子页面,在该页面可以看到上传的图片,如图 6-5 所示。进入数据库子页面,在集合 news 中可以看到刚添加的新闻记录,如图 6-6 所示。

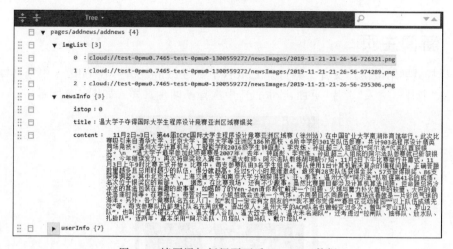

图 6-4　填写添加新闻页面后 AppData 数据

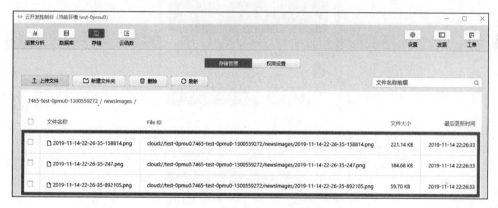

图 6-5　云开发控制台存储的图片

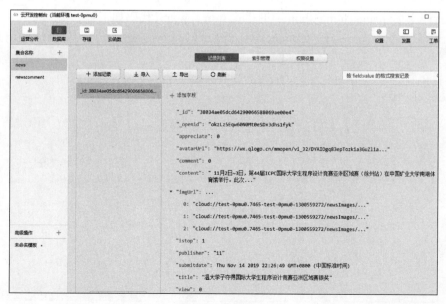

图 6-6　新闻记录

6.3 新闻主页

选择样式：用微信开发者工具打开 ColorUI 中的 demo 文件，在 tabBar 中选择"组件"，然后单击"卡片"，进入"卡片"页面，ColorUI 中选择新闻主页样式如图 6-7 所示。

图 6-7 ColorUI 中选择新闻主页样式

在 ColorUI 中打开 component/card/card.wxml 页面找到对应代码，经过适当编辑后，本案例新闻主页（pages/home/home）的样式 home.wxml 代码如下：

```
1   <cu-custom bgColor="bg-gradual-pink" isBack="{{false}}">
2     <view slot="backText">返回</view>
3     <view slot="content">新闻主页</view>
4   </cu-custom>
5
6   <block wx:for="{{newslist}}" wx:key="index" wx:for-item="item">
7     <view class="cu-card dynamic no-card">
8       <view class="cu-item shadow" data-id="{{item._id}}" bindtap="newsClick">
9         <view class="cu-list menu-avatar">
10          <view class="cu-item">
11            <view class="cu-avatar round lg"
12   style="background-image:url({{item.avatarUrl}});"></view>
13            <view class="content flex-sub">
14              <view>{{item.publisher}}</view>
15              <view class="text-gray text-sm flex justify-between">
16                {{item.time}}
```

```
17              </view>
18            </view>
19          </view>
20        </view>
21        <view class = "text-content">
22          {{item.title}}
23        </view>
24        <view class = "grid flex-sub padding-lr ">
25          <image style = "width:100%" src = "{{item.imgUrl[0]}}" mode = "aspectFill"></image>
26        </view>
27        <view class = "text-gray text-lg text-right padding">
28          <text class = "cuIcon-attentionfill margin-lr-xs text-green"></text>{{item.view}}
29          <text class = "cuIcon-appreciatefill margin-lr-xs text-cyan"></text>{{item.appreciate}}
30          <text class = "cuIcon-messagefill margin-lr-xs text-blue"></text>{{item.comment}}
31        </view>
32      </view>
33    </view>
34 </block>
```

代码第6～34行对新闻记录newslist做列表渲染,代码第8行newsClick事件执行用户单击该新闻记录,跳转到新闻详情页,newsClick事件通过代码data-id="{{item._id}}"读取新闻记录_id字段。第25行显示了新闻记录中第一张图片,在该记录的下面有3个记录,分别为浏览次数,点赞次数和评论数。添加新闻数据成功后,新闻主页显示效果如图6-8所示。

图6-8 新闻主页显示效果

读者要学会使用 data-key="{{...}}" 的方式（key 为用户自定义属性）把页面 WXML 中的数据传递给相应的 JS 文件，在 JS 文件中的事件中通过 event.currentTarget.dataset.key 获取页面传递过来的数据，例如：

```
<view bindtap="getData" data-Name="厉旭杰" data-age='30' data-cornet-phone="661780">
从页面向后台传值
</view>
```

在 getData 事件中，使用 console.log(event)。

```
getData: function (event) {
  console.log(event)
}
```

在控制台中 data-key 值的获取结果如图 6-9 所示。

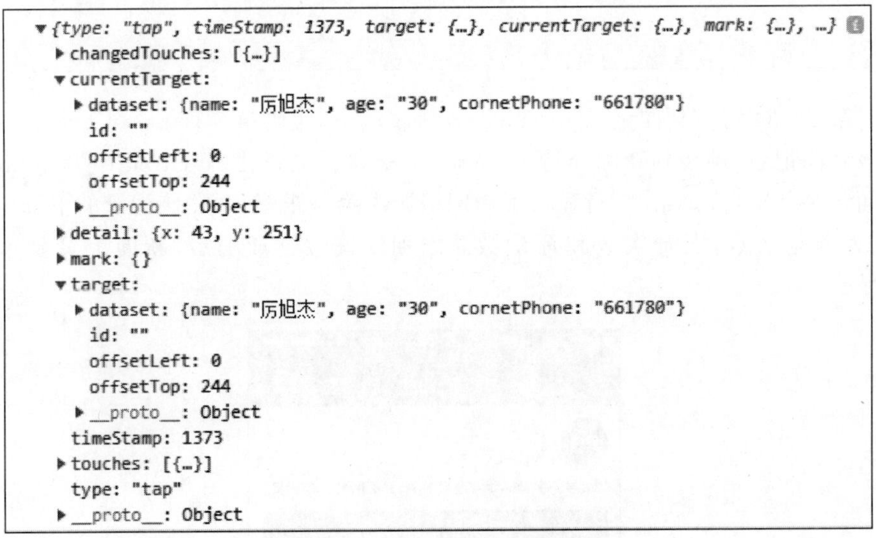

```
▼ {type: "tap", timeStamp: 1373, target: {...}, currentTarget: {...}, mark: {...}, ...}
  ▶ changedTouches: [{...}]
  ▼ currentTarget:
    ▶ dataset: {name: "厉旭杰", age: "30", cornetPhone: "661780"}
      id: ""
      offsetLeft: 0
      offsetTop: 244
    ▶ __proto__: Object
  ▶ detail: {x: 43, y: 251}
  ▶ mark: {}
  ▼ target:
    ▶ dataset: {name: "厉旭杰", age: "30", cornetPhone: "661780"}
      id: ""
      offsetLeft: 0
      offsetTop: 244
    ▶ __proto__: Object
    timeStamp: 1373
  ▶ touches: [{...}]
    type: "tap"
  ▶ __proto__: Object
```

图 6-9　控制台中 data-key 值的获取结果

这里需要注意两点：

（1）data-key 以 data-开头，多个单词由连字符"-"链接，不能有大写（大写会自动转换为小写）如 data-element-type，最终在 event.currentTarget.dataset 中会将连字符转换为驼峰 elementType（例如 cornetPhone）；

（2）target 和 currentTarget 都能获取 data-key 的值，读者要注意 target 和 currentTarget 的区别，尽量使用 currentTarget 的值。

相应的 home.js 代码如下：

```
const db = wx.cloud.database()
var util = require('../../utils/util.js')
Page({
  data: {
    newslist: {},
```

```
6    },
7
8    onLoad: function (options) {
9      db.collection('news').orderBy('istop', 'desc').orderBy('submitdate', 'desc').get().then(res =>{
10       for (var index in res.data) {
11         res.data[index].time = util.formatTime(res.data[index].submitdate)
12       }
13       this.setData({newslist: res.data})
14     })
15     .catch(err =>{
16       console.error(err)
17     })
18   },
19
20   onShow: function () {
21     this.onLoad()
22   },
23
24   newsClick: function (event) {
25     wx.cloud.callFunction({
26       name: 'news',
27       data: {
28         $url: "view_inc",      //要调用的路由的路径,传入准确路径或者通配符 *
29         id: event.currentTarget.dataset.id
30       },
31       success: res =>{
32         //console.log(res)
33       }
34     })
35     wx.navigateTo({
36       url: '../newsdetail/newsdetail?id=' + event.currentTarget.dataset.id
37     })
38   },
39 })
```

代码第8～18行查询了 news 集合中的记录,先按照是否置顶,然后按照时间降序排列。submitdate 字段为 object 类型,代码第11行把该类型转换为字符串。用户每次点击该新闻,则该新闻浏览次数需要增加1,由于数据库中 news 集合的权限设置为"所有用户可读,仅创建者可读写",因此除了该记录创建者,其他用户对该集合只有读的权限,没有写的权限,用户每次点击该新闻,浏览次数增加操作需要在云函数上执行。代码第24～38行,用户点击该新闻记录,调用云函数,执行浏览次数加1。

每条新闻记录都有浏览次数、点赞次数和评论数3个字段,因此通常需要3个云函数来分别处理,这里采用 TcbRouter,只需要一个云函数 news 来处理这3个请求,相应的云函数 news.js 代码如下:

```javascript
1   //云函数入口文件
2   const cloud = require('wx-server-sdk')
3   cloud.init()
4   const TcbRouter = require('tcb-router');
5   const db = cloud.database()
6   const _ = db.command
7   //云函数入口函数
8   exports.main = async (event, context) =>{
9
10    const app = new TcbRouter({ event });
11    //路由为字符串,该中间件只适用于 view_inc 路由
12    app.router('view_inc', async (ctx, next) =>{
13      try {
14        return await db.collection('news').doc(event.id).update({
15          //data 传入需要局部更新的数据
16          data: {
17            view: _.inc(1)
18          }
19        })
20      } catch (e) {
21        console.error(e)
22      }
23      await next();           //执行下一中间件
24    })
25
26    //路由为字符串,该中间件只适用于 appreciate_inc 路由
27    app.router('appreciate_inc', async (ctx, next) =>{
28      try {
29        return await db.collection('news').doc(event.id).update({
30          //data 传入需要局部更新的数据
31          data: {
32            appreciate: _.inc(event.value)
33          }
34        })
35      } catch (e) {
36        console.error(e)
37      }
38      await next();           //执行下一中间件
39    })
40
41    //路由为字符串,该中间件只适用于 comment_inc 路由
42    app.router('comment_inc', async (ctx, next) =>{
43      try {
44        return await db.collection('news').doc(event.id).update({
45          //data 传入需要局部更新的数据
46          data: {
47            comment: _.inc(1)
48          }
49        })
```

```
50      }catch(e){
51        console.error(e)
52      }
53      await next();  //执行下一中间件
54    })
55
56    return app.serve();
57  }
```

在 news 云函数中,采用了 3 个 app.router,分别对应 view_inc 路由、appreciate_inc 路由和 comment_inc 路由。代码第 10~24 行执行 view_inc 路由,使数据库中 view 字段增加 1;代码第 26~39 行执行 appreciate_inc 路由,用户单击新闻详情页中的"点赞"按钮,该字段增加 1,单击"取消点赞"按钮,该字段减少 1;代码第 41~54 行执行 comment_inc 路由,用户评论 1 次,则评论数增加 1。这里使用 db.command.inc 原子操作,用于指示字段自增某个值,这是个原子操作。使用这个操作指令而不是先读数据、再加、再写回的好处是:①原子性,多个用户同时写,对数据库来说都是将字段加 1,不会有后来者覆写前者的情况;②减少一次网络请求,不需要先读再写。

6.4 新闻详情页

新闻详情页(pages/newsdetail/newsdetail)的样式采用的 ColorUI 样式如图 6-7 所示,修改后的 detail.wxml 代码如下:

```
1   <cu-custom bgColor="bg-gradual-pink" isBack="{{true}}">
2     <view slot="backText">返回</view>
3     <view slot="content">新闻详情页</view>
4   </cu-custom>
5
6   <swiper class="screen-swiper {{DotStyle?'square-dot':'round-dot'}}" indicator-dots="true"
7   circular="true" autoplay="true" interval="5000" duration="500" bindtap="ViewImage">
8     <swiper-item wx:for="{{news.imgUrl}}" wx:key>
9       <image src="{{item}}" mode="aspectFill"></image>
10    </swiper-item>
11  </swiper>
12
13  <view class="cu-card dynamic no-card">
14    <view class="cu-item shadow" data-id="{{item._id}}">
15      <view class="cu-list menu-avatar">
16        <view class="cu-item">
17          <view class="cu-avatar round lg" style="background-image:url({{news.avatarUrl}});"></view>
18          <view class="content flex-sub">
19            <view>{{news.publisher}}</view>
20            <view class="text-gray text-sm flex justify-between">
21              {{news.time}}
22            </view>
```

```
23          </view>
24        </view>
25      </view>
26      <view class="text-content text-lg">
27        {{news.title}}
28      </view>
29      <view class="text-gray text-df padding">
30        <text decode="{{true}}" space="emsp">{{news.content}}</text>
31      </view>
32    </view>
33
34    <view class="cu-bar foot input {{InputBottom!=0?'cur':''}}" style="bottom:{{InputBottom}}px">
35      <view class="action">
36        <text class="text-grey">评论:</text>
37      </view>
38      <input class="solid-bottom" bindfocus="InputFocus" bindblur="InputBlur" adjust-
39 position="{{false}}" focus="{{false}}" maxlength="300" cursor-spacing="10"
40 bindinput="inputcomment" value='{{message}}'></input>
41      <view class="action">
42        <text class="cuIcon-emojifill text-grey"></text>
43        <text class="cuIcon-{{fill?'appreciatefill':'appreciate'}} {{fill?'text-pink':
44 'text-grey'}}" bindtap="appreciate"></text>
45      </view>
46      <button class="cu-btn bg-green shadow" bindtap="sendmessage">发送</button>
47    </view>
48 </view>
49
50 <view class="cu-card dynamic no-card" style="margin-bottom:60px">
51    <view class="cu-item shadow">
52      <view class="cu-list menu-avatar comment solids-top">
53        <block wx:for="{{comment}}" wx:key="index" wx:for-item="item">
54          <view class="cu-item">
55            <view class="cu-avatar round" style="background-image:url({{item.avatarUrl}});"></view>
56            <view class="content">
57              <view class="text-grey">昵称:{{item.nickName}}</view>
58              <view class="text-gray text-content text-df">
59                性别:<block wx:if="{{item.gender==1}}">男</block>
60                <block wx:else>女</block>
61              </view>
62              <view class="bg-grey padding-sm radius margin-top-sm text-sm">
63                <view class="flex">
64                  <view class="flex-sub">{{item.message}}</view>
65                </view>
66              </view>
67              <view class="margin-top-sm flex justify-between">
68                <view class="text-gray text-df">{{item.time}}</view>
69                <view>
70                  <text class="cuIcon-appreciatefill text-red"></text>
71                  <text class="cuIcon-messagefill text-gray margin-left-sm"></text>
```

```
72                    </view>
73                  </view>
74                </view>
75              </view>
76        </block>
77      </view>
78    </view>
79 </view>
```

代码第 6~11 行使用轮播图播放新闻图片,用户点击轮播图上的图片可以进行全屏浏览。代码第 13~32 行显示新闻信息,包括发布人头像、发布人昵称、新闻发布时间、新闻标题和新闻内容,需要注意的是,新闻内容里包含了空格和换行\n,必须使用< text decode="{{true}}" space="emsp">{{news.content}}</text>把新闻内容放在< text >< /text >标签中才能正常显示空格和换行。代码第 34~48 行显示了最下方的评论组件,需要说明的是,微信开发者工具中,在评论组件中输入评论,不会弹出手机键盘,读者要看输入评论时,弹出手机键盘效果,可以使用真机调试,用微信扫描弹出的二维码中可以进行真机测试。代码第 50~79 行显示了用户的评论。新闻详情页最终的显示效果如图 6-10 所示。

图 6-10 新闻详情页最终的显示效果

相应的 newsdetail.js 代码如下:

```
1   var util = require('../../utils/util.js')
2   const db = wx.cloud.database()
3   const app = getApp();
```

```
4   Page({
5     data: {
6       news: {},
7       comment: {},
8       message: '',
9       newsid: '',
10      fill: false
11    },
12
13    onLoad: function(options) {
14      const newsid = options.id
15      db.collection('news').doc(newsid).get().then(res =>{
16        res.data.time = util.formatTime(res.data.submitdate)
17        this.setData({
18          newsid: newsid,
19          news: res.data
20        })
21      })
22
23      db.collection('newscomment').where({
24        newsid: newsid
25      }).orderBy('submitdate', 'desc').get().then(res =>{
26        for (var index in res.data)
27          res.data[index].time = util.formatTime(res.data[index].submitdate)
28        this.setData({
29          comment: res.data
30        })
31      })
32    },
33
34    ViewImage(e) {
35      wx.previewImage({
36        urls: this.data.news.imgUrl,
37        current: e.currentTarget.dataset.url
38      });
39    },
40
41    InputFocus(e) {
42      this.setData({
43        InputBottom: e.detail.height
44      })
45    },
46
47    InputBlur(e) {
48      this.setData({
49        InputBottom: 0
50      })
51    },
52
```

```
53    inputcomment(event) {
54      this.setData({
55        message: event.detail.value
56      })
57    },
58
59    appreciate: function(event) {
60      console.log('fill')
61      var fill = this.data.fill;
62      var value = fill ? -1 : 1
63      wx.cloud.callFunction({
64        name: 'news',
65        data: {
66          $url: "appreciate_inc",          //要调用的路由的路径,传入准确路径或者通配符 *
67          id: this.data.newsid,
68          value: value
69        },
70        success: res =>{
71          //console.log(res)
72        }
73      })
74      this.setData({
75        fill: !fill
76      })
77    },
78
79    sendmessage(event) {
80      var userInfo = wx.getStorageSync('userInfo');
81      if (userInfo == "") {
82        wx.navigateTo({
83          url: '../auth/auth'
84        })
85        return
86      }
87      var newsid = this.data.newsid
88      db.collection('newscomment').add({
89        data: {
90          newsid: newsid,
91          message: this.data.message,
92          nickName: userInfo.nickName,
93          gender: userInfo.gender,
94          avatarUrl: userInfo.avatarUrl,
95          submitdate: db.serverDate(),
96        }
97      }).then(res =>{
98        db.collection('newscomment').where({
99          newsid: newsid
100       }).orderBy('submitdate', 'desc').get().then(res =>{
101         for (var index in res.data)
```

```
102              res.data[index].time = util.formatTime(res.data[index].submitdate)
103            this.setData({
104              comment: res.data, message:''
105            })
106          })
107          wx.cloud.callFunction({
108            name: 'news',
109            data: {
110              $url: "comment_inc", //要调用的路由的路径
111              id: newsid,
112            },
113            success: res =>{
114              console.log(res)
115            }
116          })
117          wx.showToast({
118            title: '评论成功...',
119            icon: 'none',
120            duration: 1500
121          })
122        })
123      },
124    })
```

代码第 13~32 行，在 onLoad 加载事件中，读取数据库集合 news 中 _id 为 newsid 的新闻记录，并读取数据库 newscomment 集合中 newsid 字段为 newsid 的记录。代码第 34~39 行执行当用户点击轮播图上的图片时对新闻图片进行全屏浏览。代码第 41~51 行执行焦点在评论输入框中和不在输入框时，评论组件离页面底部的位置，该效果只能在真机中才能看到。代码第 53~57 行读取用户输入的评论。代码第 59~77 行执行用户点赞操作，用户可以对该新闻进行点赞，也可以在点赞后取消点赞，点赞数的操作在云函数中执行。代码第 79~123 行执行发布用户评论操作，发布评论之前，用户先要进行登录授权，代码第 80~86 行检查用户是否登录授权，代码第 88~123 行先通过 collection.add 把用户评论添加进数据库 newscomment 集合中，添加成功后读取 newscomment 集合中该新闻的评论信息，因为可能有多个用户对该新闻进行评论，用户评论后不能简单地把自己的评论显示在页面上，需要再次查询数据库 newscomment 集合中当前时间所有该新闻的评论，最后执行 news 集合中该新闻评论数增加 1。云开发控制台数据库 newscomment 集合的数据如图 6-11 所示。

需要说明的是，本页面演示了用户对新闻进行评论的功能，如果你的微信小程序需要上线，需要仔细阅读微信小程序开发文档中"运营"→"开放的服务类目"（https://developers.weixin.qq.com/miniprogram/product/material/），微信对个人主体和非个人主体开放的服务类目进行了详细说明。对于个人用户的微信小程序，普通用户无法发布新闻和对新闻进行评论交流。

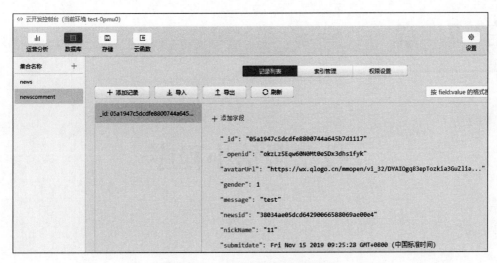

图 6-11　newscomment 集合的数据

第 7 章

投票微信小程序

投票微信小程序包含 4 个页面：

```
1  "pages":[
2    "pages/auth/auth",              //授权页面
3    "pages/addvote/addvote",        //添加投票页面
4    "pages/home/home",              //投票主页
5    "pages/votedetail/votedetail",  //投票页面
6  ],
```

其中，包含 2 个数据库集合 vote 和 ticket，权限均为"所有用户可读，仅创建者可读写"。

7.1 授权页面

授权页面和新闻微信小程序授权页面类似，唯一不同的是在投票的时候为了区分哪个用户进行了投票，需要获取用户的唯一标识符 openid。

授权页面（pages/auth/auth）的样式 auth.wxml 如下：

```
1   <cu-custom bgColor="bg-gradual-pink" isBack="{{false}}">
2     <view slot="content">登录</view>
3   </cu-custom>
4
5   <view class="bg-white solid-bottom">
6     <view class="title">
7       <open-data type="userAvatarUrl"></open-data>
8     </view>
9   </view>
10  <view class="padding bg-white">
11    <view class="text-black">该应用将获取以下授权:</view>
12    <view class="text-grey">获取您的公开信息(昵称,头像,性别等)</view>
```

```
13      </view>
14      <view style = 'display: flex; justify-content: center; bg-white'>
15      <button bindtap = "refuseAuth" class = "cu-btn xl bg-green padding-sm margin-xs">拒绝</button>
16      <button bindgetuserinfo = "bindGetUserInfo" open-type = 'getUserInfo' class = "cu-btn xl
17      bg-green padding-sm margin-xs">允许</button>
18      </view>
```

授权页面的显示效果如图 7-1 所示。

图 7-1　投票微信小程序授权页面的显示效果

相应的 auth.js 代码如下：

```
1     Page({
2       refuseAuth: function () {
3         wx.switchTab({
4           url: '../home/home'
5         })
6       },
7       bindGetUserInfo: function (e) {
8         if (e.detail.userInfo) {
9           wx.cloud.callFunction({
10            name: 'login',
11            data: {},
12            success: res =>{
13              e.detail.userInfo.openid = res.result.openid
14              wx.setStorageSync('userInfo', e.detail.userInfo);
15              wx.showToast({
```

```
16                title: '授权成功...',
17                icon: 'none',
18                duration: 1500
19              })
20              wx.navigateBack({
21                delta: 1
22              })
23            }
24          })
25        }
26    },
27  })
```

代码第 2~6 行对应用户单击"拒绝授权"按钮，返回上一页面。代码第 7~26 行对应用户单击"允许授权"按钮执行代码，其中第 9~24 行调用云函数 login 获取用户 openid。

云函数 login 代码如下：

```
1   const cloud = require('wx-server-sdk')
2   cloud.init()
3   exports.main = async (event, context) =>{
4     const wxContext = cloud.getWXContext()
5
6     return {
7       event,
8       openid: wxContext.OPENID,
9       appid: wxContext.APPID,
10      unionid: wxContext.UNIONID,
11    }
12  }
```

在云函数目录下新建云函数 login，系统会自动生成云函数代码，这里的 login 云函数不需要修改代码，用系统自动生成的代码就可以获取用户的 openid。其中第 8 行返回用户的 openid。用户授权登录后 Storage 缓存的数据如图 7-2 所示。

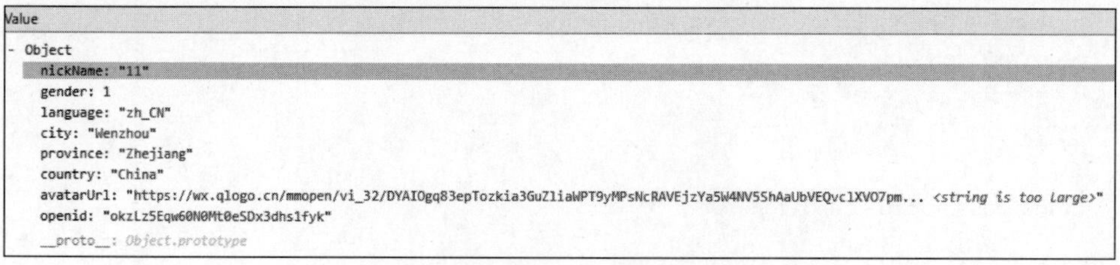

```
Value
- Object
    nickName: "11"
    gender: 1
    language: "zh_CN"
    city: "Wenzhou"
    province: "Zhejiang"
    country: "China"
    avatarUrl: "https://wx.qlogo.cn/mmopen/vi_32/DYAI0gq83epTozkia3GuZ1iaWPT9yMPsNcRAVEjzYa5W4NV5ShAaUbVEQvc1XVO7pm... <string is too large>"
    openid: "okzLz5Eqw60N0Mt0eSOx3dhs1fyk"
    __proto__: Object.prototype
```

图 7-2　用户授权登录后 Storage 缓存的数据

7.2　添加投票页面

添加投票页面（pages/addvote/addvote）样式采用 ColorUI 中的表单样式（代码见 pages/component/form），addvote.wxml 代码如下：

```
1   <cu-custom bgColor="bg-gradual-pink" isBack="{{true}}">
2     <view slot="backText">返回</view>
3     <view slot="content">添加投票</view>
4   </cu-custom>
5
6   <form>
7     <view class="cu-form-group">
8       <input placeholder="投票标题" bindinput="inputtitle"></input>
9     </view>
10    <view class="cu-form-group">
11      <input placeholder="补充描述(选填)" bindinput="inputdescription"></input>
12    </view>
13    <view class="cu-form-group">
14      <view class="title">投票类型</view>
15      <picker bindchange="PickerChange" value="{{index}}" range="{{picker}}">
16        <view class="picker">
17          {{picker[index]}}
18        </view>
19      </picker>
20    </view>
21    <block wx:for="{{voteoptions}}" wx:key>
22      <view class="cu-form-group">
23        <text class="cuIcon-roundclosefill text-red" style="width:50rpx" data-id="{{index}}"
24  bindtap="minusoptions"></text>
25        <input placeholder="选项" data-id="{{index}}" bindinput="inputoptions" value=
26  '{{item.option}}'></input>
27      </view>
28    </block>
29    <view class="cu-form-group" bindtap="addoptions">
30      <text class="cuIcon-roundaddfill text-blue" style="width:50rpx"></text>
31      <input placeholder-class="text-blue" disabled placeholder="添加选项"></input>
32    </view>
33    <view class="cu-form-group margin-top">
34      <view class="title">截止日期</view>
35      <picker mode="date" value="{{date}}" start="2015-09-01" end="2020-09-01"
36  bindchange="DateChange">
37        <view class="picker">
38          {{date}}
39        </view>
40      </picker>
41    </view>
42    <view class="cu-form-group">
43      <view class="title">匿名投票</view>
44      <switch class="red sm" checked="{{announymous}}" bindchange="switchchange"></switch>
45    </view>
46  </form>
47  <view class="padding-xl">
48    <button class="cu-btn block bg-blue margin-tb-sm lg" bindtap="submit">完成</button>
49  </view></view>
```

代码第 21～32 行是添加投票选项的功能，这部分是整个添加投票页面的难点，因为投票的选项数量往往是不固定的，所以管理员可以增加投票选项和删除投票选项。该页面目前只支持投票类型为单选。添加投票页面的效果如图 7-3 所示。

图 7-3　添加投票页面的效果

相应的 addvote.js 代码如下：

```
1   const db = wx.cloud.database()
2   var util = require('../../utils/util.js')
3   Page({
4     data:{
5       title:'',
6       description:'',
7       voteoptions:[{
8         option:''
9       },
10      {
11        option:''
12      }
13      ],
14      optionnum:2,
15      votetype:'单选',
16      date:'2019-11-15',
17      picker:['单选'],
18      index:0,
19      announymous:true
20    },
```

```
21
22      onShow: function() {
23        var userInfo = wx.getStorageSync('userInfo');
24        if (userInfo == "") {
25          wx.navigateTo({
26            url: '../auth/auth'
27          })
28        } else {
29          this.setData({
30            userInfo: userInfo
31          })
32          var todaydate = util.formatDatastring(new Date())
33          this.setData({
34            date: todaydate
35          })
36        }
37      },
38
39      addoptions: function(event) {
40        var optionnum = this.data.optionnum
41        var voteoptions = this.data.voteoptions
42        var voteoption1 = {}
43        voteoption1.option = ""
44        voteoptions[optionnum] = voteoption1
45        this.setData({
46          voteoptions: voteoptions,
47          optionnum: optionnum + 1
48        })
49      },
50
51      minusoptions: function(event) {
52        console.log(event.currentTarget.dataset.id)
53        var id = event.currentTarget.dataset.id
54        var optionnum = this.data.optionnum
55        var voteoptions = this.data.voteoptions
56        voteoptions.splice(id, 1)
57        this.setData({
58          voteoptions: voteoptions,
59          optionnum: optionnum - 1
60        })
61      },
62
63      inputtitle: function(event) {
64        this.setData({
65          title: event.detail.value
66        })
67      },
68      inputdescription: function(event) {
69        this.setData({
```

```
70          description: event.detail.value
71        })
72    },
73
74    inputoptions: function(event) {
75      var id = event.currentTarget.dataset.id
76      var buff = event.detail.value
77      var voteoptions = this.data.voteoptions
78      voteoptions[id].option = buff
79      console.log(voteoptions)
80      this.setData({
81        voteoptions: voteoptions
82      })
83    },
84
85    PickerChange(e) {
86      console.log(e);
87      var picker = this.data.picker
88      var index = e.detail.value
89      this.setData({
90        index: index,
91        votetype: picker[index]
92      })
93    },
94
95    DateChange(e) {
96      this.setData({
97        date: e.detail.value
98      })
99    },
100
101   switchchange: function(event) {
102     this.setData({
103       announymous: event.detail.value
104     })
105   },
106
107   submit: function(event) {
108     let userInfo = this.data.userInfo
109     var date = this.data.date
110     var time = this.data.time
111     var deadline = new Date(date + " " + '23:00')
112     var optionname = ['A', 'B', 'C', 'D', 'E', 'F']
113     var voteoptions = this.data.voteoptions
114     if (voteoptions.length > 6) {
115       wx.showToast({
116         title: '选项不能超过6个',
117         icon: 'success',
118         duration: 2000
```

```
119          })
120          return
121        } else {
122          for (var index in voteoptions) {
123            voteoptions[index].name = optionname[index]
124            voteoptions[index].num = 0
125          }
126          db.collection('vote').add({
127            data: {
128              publisher: userInfo.nickName,
129              avatarUrl: userInfo.avatarUrl,
130              title: this.data.title,
131              description: this.data.description,
132              voteoptions: voteoptions,
133              votetype: this.data.votetype,
134              submitdate: db.serverDate(),
135              deadline: deadline,
136              announymous: this.data.announymous,
137              available: true
138            }
139          })
140          .then(res =>{
141            wx.showToast({
142              title: '投票添加成功!',
143              icon: 'success',
144              duration: 2000
145            })
146            wx.reLaunch({
147              url: '../home/home'
148            })
149          })
150        }
151      }
152  })
```

代码第 4～20 行对应 data 数据项，title 为投票的标题，description 为投票的补充描述，voteoptions 为投票的选项内容，optionnum 为选项的数量，初始值为 2，votetype 为投票类型，这里只支持单选，date 为投票截止日期初始值，announymous 为投票是否支持匿名投票。因为 switchTab 切换页面的时候不会调用 onLoad 事件，所以数据加载放在 onShow 事件中，添加投票需要管理员权限，因此在 onShow 事件中需要检查用户是否授权登录，如果没有授权登录，则跳转到授权页面，在 onShow 事件中本项目设置投票截止时间为当天。代码第 39～49 行实现单击添加选项，投票的选项增加一项，投票选项存储在 voteoptions 数组中。代码第 51～61 行对应单击选项前的"×"图标，实现删除该投票选项，也就是从数组 voteoptions 中删除该选项数据。代码第 63～67 行获取投票标题。代码第 68～72 行获取投票的补充描述。代码第 74～83 行获取投票选项的内容。代码第 85～93 行获取投票的类型。代码第 95～99 行获取截止日期。代码第 101～105 行获取是否匿名投票。代码第 107～

151 行实现提交投票,其中代码第 111 行设置投票截止日期的时间默认为 23:00。代码第 114 行检查投票选项是否超过 6 项,代码第 122～125 行设置每个选项的投票数为 0,代码第 126～139 行实现把投票数据写入数据库集合 vote 中。userInfo.nickName 和 userInfo.avatarUrl 分别为微信用户的昵称和头像,db.serverDate() 为添加投票的服务器时间,available 这个字段为投票是否在有效时间内,如果该字段为 true,则表示正在进行的投票;若该字段为 false,则表示已经结束的投票。填写添加投票页面后,AppData 中的数据如图 7-4 所示。数据添加成功后,数据库的投票数据显示如图 7-5 所示。

图 7-4　添加投票页面 AppData 中的数据

图 7-5　数据库的投票数据显示

7.3 投票主页

投票主页(pages/home/home)的导航栏样式选择：用微信开发者工具打开 ColorUI 中的 demo 文件，在 tabBar 中选择"组件"，然后单击"导航栏"，进入"导航栏"页面，对应 ColorUI 项目中(pages/componet/nav)页面；ColorUI 中导航栏样式如图 7-6 所示。

图 7-6　ColorUI 中导航栏样式

投票主页 home.wxml 代码为：

```
1   <cu-custom bgColor="bg-gradual-pink" isBack="{{false}}">
2     <view slot="backText">返回</view>
3     <view slot="content">投票</view>
4   </cu-custom>
5
6   <scroll-view scroll-x class="bg-white nav">
7     <view class="flex text-center">
8       <view class="cu-item flex-sub {{index==TabCur?'text-orange cur':''}}" wx:for="
9   {{navTab}}" wx:key bindtap="tabSelect" data-id="{{index}}">
10        {{item}}
11      </view>
12    </view>
13  </scroll-view>
14
15  <view class="cu-list menu sm-border">
16    <block wx:for="{{votelist}}" wx:key="index">
```

```
17        <view class = "cu-item">
18          <image style = "width: 50px; height: 50px" src = '/images/toupiao.png'></image>
19          <view class = "content">
20            <view class = "text-df text-blue">投票主题:{{item.title}}</view>
21            <view class = "text-df text-blue">投票开始时间:{{item.time}}</view>
22            <view class = "text-df text-blue">投票结束时间:{{item.deadtime}}</view>
23          </view>
24          <view class = "action">
25            <view style = 'flex-basis: 15%; justify-content:center;' bindtap = "gotovote"
26  data-index = '{{index}}' data-id = '{{item._id}}'>
27              <button class = "cu-btn round bg-green ">{{item.btnText}}</button>
28          </view>
29        </view>
30      </view>
31    </block>
32  </view>
```

代码第6～13行对应导航栏,导航栏分为两栏,分别为"正在进行的投票"和"已经结束的投票";代码第15～32行显示投票信息。投票主页的显示效果如图7-7所示。

图 7-7　投票主页的显示效果

相应的home.js代码如下:

```
1  const db = wx.cloud.database()
2  var util = require('../../utils/util.js')
3  Page({
4    data: {
```

```
5        TabCur: 0,
6        navTab: ['正在进行的投票', '已经结束的投票'],
7        votelist: [],
8      },
9
10     LoadAvailablevote(){
11       wx.showLoading({
12         title: '数据加载中...',
13       })
14       db.collection('vote').where({
15         available: true
16       }).orderBy('submitdate', 'desc')
17         .get().then(res =>{
18           for (var index in res.data) {
19             res.data[index].time = util.formatTime(res.data[index].submitdate)
20             res.data[index].deadtime = util.formatTime(res.data[index].deadline)
21             res.data[index].btnText = '投票'
22           }
23           this.setData({
24             votelist: res.data
25           })
26           wx.hideLoading()
27         })
28     },
29
30     LoadUnavailablevote() {
31       wx.showLoading({
32         title: '数据加载中...',
33       })
34       db.collection('vote').where({
35         available: false
36       }).orderBy('submitdate', 'desc')
37         .get().then(res =>{
38           for (var index in res.data) {
39             res.data[index].time = util.formatTime(res.data[index].submitdate)
40             res.data[index].deadtime = util.formatTime(res.data[index].deadline)
41             res.data[index].btnText = '查看'
42           }
43           this.setData({
44             votelist: res.data
45           })
46           wx.hideLoading()
47         })
48     },
49
50     onShow: function () {
51       if (this.data.TabCur == 0) {
52         this.LoadAvailablevote()
54       } else if (this.data.TabCur == 1) {
```

```
55          this.LoadUnavailablevote()
56        }
57      },
58
59      tabSelect(e) {
60        this.setData({
61          TabCur: e.currentTarget.dataset.id
62        })
63        if (this.data.TabCur == 0) {
64          this.LoadAvailablevote()
65
66        } else if (this.data.TabCur == 1) {
67          this.LoadUnavailablevote()
68        }
69      },
70
71      gotovote: function (event) {
72        var userInfo = wx.getStorageSync('userInfo');
73        if (userInfo == "") {
74          wx.navigateTo({
75            url: '../auth/auth'
76          })
77          return
78        }
79        wx.navigateTo({
80          url: '../votedetail/votedetail? id=' + event.currentTarget.dataset.id
81        })
82      }
83    })
```

本页面导航栏分为"正在进行的投票"和"已经结束的投票",其中代码第 10～28 行加载正在进行的投票,也就是查询数据库 vote 集合中 available 字段为 true 的记录;代码第 30～48 行加载已经结束的投票,也就是查询数据库 vote 集合中 available 字段为 false 的记录。代码第 50～57 行在 onShow 事件根据不同的导航栏,加载不同的投票记录。代码第 59～69 行对应导航栏切换事件。代码第 71～82 行对应单击"投票"按钮事件,如果用户未授权登录则跳转到授权页面,如果已经登录,则跳转到投票页面。

7.4 投票页面

投票页面(pages/votedetail/votedetail)样式中含有进度条,进度条的样式选用:用微信开发者工具打开 ColorUI 中的 demo 文件,在 tabBar 中选择"元素",然后单击"进度条",进入"进度条"页面,对应 ColorUI 项目中(pages/basics/progress)页面。ColorUI 中进度条样式如图 7-8 所示。

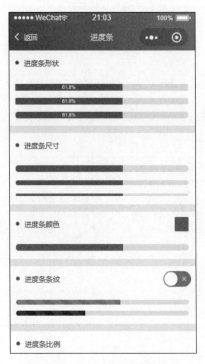

图 7-8　ColorUI 中进度条样式

投票页面 votedetail.js 代码如下：

```
1   <cu-custom bgColor = "bg-gradual-pink" isBack = "{{true}}">
2     <view slot = "backText">返回</view>
3     <view slot = "content">投票页面</view>
4   </cu-custom>
5
6   <view class = "cu-bar bg-white solid-bottom">
7     <view class = "action">
8       <text class = "cuIcon-title text-orange"></text>[{{vote.votetype}}]{{vote.title}}
9     </view>
10  </view>
11  <view class = "cu-bar bg-white solid-bottom">
12    <view class = "action">
13      [问题补充]{{vote.description}}
14    </view>
15  </view>
16  <form class = "cu-bar bg-white solid-bottom">
17    <radio-group class = "radio-group" bindchange = "radioChange">
18      <label class = "radio" wx:for = "{{vote.voteoptions}}" wx:key = "index">
19        <view class = "cu-bar bg-white">
20          <view class = "action">
21            <radio value = "{{index}}" disabled = "{{isvote}}"
22              checked = "{{index == selectvote? true:false}}" />{{item.name}}.{{item.option}}
23          </view>
24        </view>
25      </label>
```

```
26      </radio-group>
27    </form>
28
29    <block wx:if = "{{isvote}}">
30      <view class = "cu-bar bg-white solid-bottom margin-top">
31        <view class = "action">
32          <text class = "cuIcon-title text-blue"></text>投票结果:
33        </view>
34      </view>
35      <view class = "padding bg-white">
36        <block wx:for = "{{vote.voteoptions}}" wx:key>
37          <view class = "cu-bar bg-white solid-bottom">选项{{item.name}}:{{item.num}} 票</view>
38          <view class = "cu-progress radius striped active">
39            <view class = "bg-red" style = "width:{{item.percent}};">{{item.percent}}</view>
40          </view>
41        </block>
42      </view>
43    </block>
44    <view class = "padding-xl">
45      <button class = "cu-btn block bg-blue margin-tb-sm lg" disabled = "{{isvote}}" bindtap =
46  "submit">提交</button>
47    </view>
```

代码第 6~27 显示投票信息,核心部分为代码第 17~26 行,使用 radio-group 单项选择器,由多个 radio 组成。代码第 29~43 行显示投票结果,包括具体每个选项的票数以及百分比。投票页面的效果如图 7-9 所示。

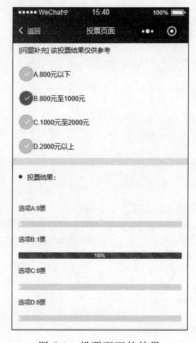

图 7-9　投票页面的效果

相应的 votedetail.js 代码如下：

```js
const db = wx.cloud.database()
Page({
  data: {
    vote: {},
    selectvote: -1,
    isvote: false,
  },

  caculatevote: function() {
    db.collection('vote').doc(this.data.vote._id).get().then(optres =>{
      var vote = optres.data
      console.log(vote)
      var total = 0
      for (var i = 0; i< vote.voteoptions.length; i++) {
        total = total + vote.voteoptions[i].num
      }
      for (var i = 0; i< vote.voteoptions.length; i++) {
        var percent = Math.round(vote.voteoptions[i].num / (total) * 10000) / 100.00 + "%"
        vote.voteoptions[i].percent = percent
      }
      this.setData({
        vote: vote
      })
    })
  },

  onLoad: function(options) {
    var id = options.id
    var userInfo = wx.getStorageSync('userInfo');
    db.collection('vote').doc(id).get().then(res =>{
      var vote = res.data
      this.setData({
        vote: vote
      })
      db.collection('ticket').where({
        id: vote._id,
        _openid: userInfo.openid
      }).get().then(ticketres =>{
        this.caculatevote()
        if (ticketres.data.length != 0) {
          this.setData({
            isvote: true,
            selectvote: ticketres.data[0].selectvote
          })
        } else {
          this.setData({
            isvote: !vote.available,
          })
```

```
49          }
50        })
51      })
52    },
53    radioChange: function(event) {
54      var selectvote = event.detail.value
55      this.setData({
56        selectvote: selectvote
57      })
58    },
59
60    submit: function(event) {
61      var userInfo = wx.getStorageSync('userInfo');
62      if (this.data.selectvote != '') {
63        db.collection('ticket').add({
64          data: {
65            id: this.data.vote._id,
66            selectvote: this.data.selectvote
67          }
68        })
69        .then(res =>{
70          wx.cloud.callFunction({
71            name: 'vote',
72            data: {
73              _id: this.data.vote._id,
74              selectvote: this.data.selectvote
75            },
76            success: res =>{
77              this.caculatevote()
78              this.setData({
79                isvote: true,
80              })
81            }
82          })
83        })
84        .catch(console.error)
85      } else {
86        wx.showToast({
87          title: '请输入选项...',
88          icon: 'success',
89          duration: 2000
90        })
91      }
92    },
93  })
```

代码第 27～52 行对应页面加载事件，其中代码第 30～34 行从数据库集合 vote 中搜索关键字为 id 的投票记录，随后代码第 35～50 行判断数据库集合 ticket 中是否存 id 字段为该投票的 id，openid 字段为当前用户 openid 的记录，如果存在记录，则表示用户已经投过票了；如果不存在记录，则表示用户还没有投过票。ticketres.data.length！= 0 表示已经投票，当前投票的选项为 ticketres.data[0].selectvote。代码第 53~58 行对应 radio-group 的响应事件，获得用户选择的选项。代码第 60~92 行对应用户提交投票结果，其中第 63~68 行往数据库集合 ticket 中插入用户投票的结果，根据用户投票的选项，需要在 vote 集合中增加相应选项的票数，因为需要修改 vote 集合，所以必须采用云函数实现。代码第 70~83 行调用云函数 vote 实现相应投票选项票数增加 1。

这里云函数 vote 的实现如下：

```
1  const cloud = require('wx-server-sdk')
2  cloud.init()
3  const db = cloud.database()
4  const _ = db.command
5  exports.main = async (event, context) =>{
6    var opt = event.selectvote
7    var options = 'voteoptions.' + opt + '.num'
8    try {
9      return await db.collection('vote').doc(event._id).update({
10       data: {
11         [options]: _.inc(1)
12       }
13     })
14   } catch (e) {
15     console.error(e)
16   }
17 }
```

在这里相应票数增加采用原子操作 db.command.inc，难点在于票数都在 voteoptions 字段中，而 voteoptions 本身是个数组，对于更新数组元素的操作读者可以参考 3.5.6 节，这里使用"点表示法"，例如 voteoptions.1.num 表示 voteoptions 数组中第一个元素中的 num 的值。在这里读者需要注意，data 中的字段名是变量，因此需要把变量放入[]中。云函数执行完以后 vote 集合中相应的选项 num 会增加 1，如图 7-10 所示。

到此投票微信小程序的 4 个页面都已经完成了，这里还有一个很重要的工作就是本项目在添加投票页面的时候设置了投票的截止日期，那么如何让投票到了截止时间后，投票的 available 字段变成 false 呢？这里就要用到云函数的定时触发器，在项目中新建云函数 runner，并添加 config.json 文件，代码如下：

```
1  const cloud = require('wx-server-sdk')
2  cloud.init()
3  const db = cloud.database()
4  const _ = db.command
5  {
6    "triggers":[
7      {
8        "name":"myTrigger",
9        "type":"timer",
```

```
10            "config":"0 0 23 * * * *"
11       }
12   ]
13 }
```

```
+ 添加字段
    "_id": "38034ae05dd6354c01d7735e5cf0170f"
    "_openid": "okzLz5Eqw60N0Mt0eSDx3dhs1fyk"
    "announymous": true
    "available": true
    "avatarUrl": "https://wx.qlogo.cn/mmopen/vi_32/DYAIOgq83epTozkia3GuZlia..."
    "deadline": Thu Nov 21 2019 23:00:00 GMT+0800 (中国标准时间)
    "description": "该投票结果仅供参考" (string)
    "publisher": "11"
    "submitdate": Thu Nov 21 2019 14:57:16 GMT+0800 (中国标准时间)
    "title": "你平均一个月用多少生活费"
  ▼ "voteoptions": ...
    ▶ 0: {"name":"A","num":0,"option":"800元以下"}
    ▶ 1: {"name":"B","num":1,"option":"800元至1000元"}
    ▶ 2: {"name":"C","num":0,"option":"1000元至2000元"}
    ▶ 3: {"name":"D","num":0,"option":"2000元以上"}
    "votetype": "单选"
```

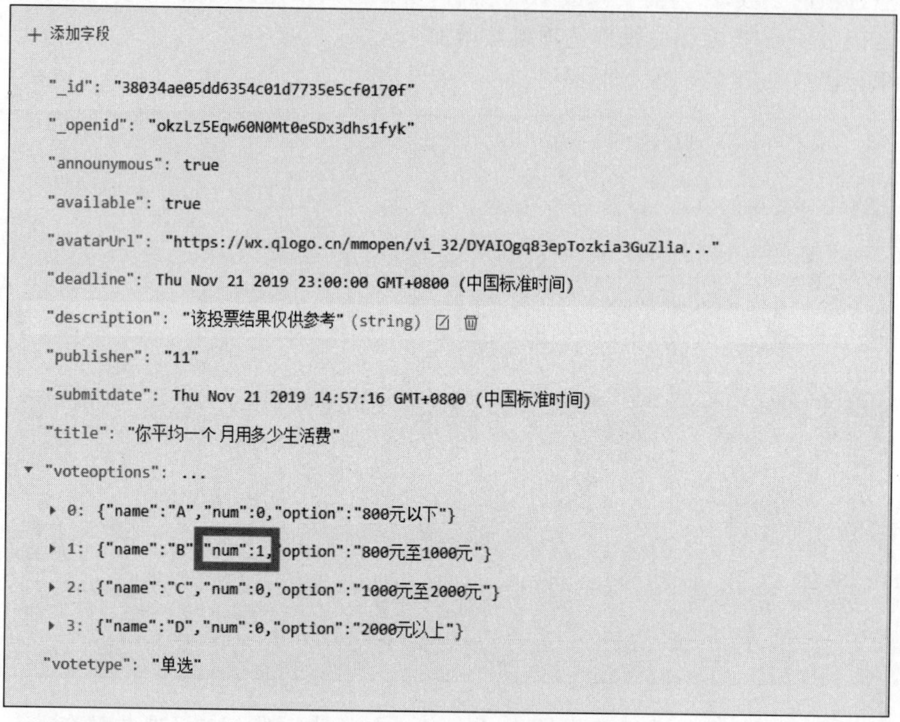

图 7-10　vote 集合选项票数增加 1

该触发器的 config 字段表示每天的 23 点就会触发,设置完触发器需要将其上传到云服务器上,相应的云函数为:

```
1  const cloud = require('wx-server-sdk')
2  cloud.init()
3  const db = cloud.database()
4
5  exports.main = async (event, context) =>{
6    let tasks = await db.collection('vote')
7      .where({
8        available: true
9      }).get()
10   let todaydata = new Date()
11   for(i = 0; i<tasks.data.length; i++)
12   {
13     if (tasks.data[i].deadline <= todaydata)
14     {
15       await db.collection('vote').doc(tasks.data[i]._id).update({
16         data:{
```

```
17            available: false
18          }
19        })
20      }
21    }
22 }
```

代码第 6~9 行查询数据库 vote 集合中所有正在进行的投票记录，随后代码第 11~21 行逐条检查正在进行的投票记录是否已经超过截止时间，如果正在进行的投票记录已经超过截止时间，则修改该投票记录 available 字段的值为 false。为了降低云函数的调用次数，本项目的触发器设置每天 23 点检查一次。

本项目的投票微信小程序和答题类微信小程序（选择题、判断题）类似，读者学习本章课程后，可以编写一个答题类微信小程序。

第 8 章

通讯录微信小程序

通讯录微信小程序包含 4 个页面：

```
1  "pages": [
2    "pages/home/home",              //项目主页
3    "pages/contact/contact",        //通讯录页面
4    "pages/deleteperson/deleteperson"  //删除人员页面
5    "pages/addperson/addperson",    //添加人员页面
6  ],
```

其中，包含 2 个数据库集合 appbox 和 contact，权限均为"所有用户可读，仅创建者可读写"。

8.1 项目主页

项目主页样式采用 ColorUI 宫格列表样式（代码见 pages/component/list），如图 8-1 所示。

本项目主页 home.wxml 样式如下：

```
1   <cu-custom bgImage="/images/sylb2244.jpg" isBack="{{false}}">
2     <view slot="content">应用程序</view>
3   </cu-custom>
4
5   <view class="cu-bar bg-white solid-bottom">
6     <view class="action">
7       <text class="cuIcon-title text-orange"></text>应用列表
8     </view>
9   </view>
10
```

```
11    <view class = "cu - list grid col - 3 ">
12      <view class = "cu - item" wx:for = "{{boxlist}}" wx:key = "index" wx:for - item = "item">
13        <view data - id = '{{item.name}}' bindtap = 'enterapplication'>
14          <image style = "width: 40px; height: 40px" src = '{{item.url}}'></image>
15          <text>{{item.name}}</text>
16        </view>
17      </view>
18    </view>
```

图 8-1 ColorUI 宫格列表样式

代码第 1～3 行对应项目头部导航栏，这里为了方便，图片 sylb2244.jpg 放在了项目本地文件夹下，读者应该把该图片放到云存储上。代码第 11～18 行采用了 ColorUI 中的宫格列表样式。通讯录项目主页显示效果如图 8-2 所示。

相应的 home.js 代码如下：

```
1   const db = wx.cloud.database()
2   Page({
3     data: {
4       boxlist: {},
5     },
6
7     onLoad: function (options) {
8       db.collection('appbox').orderBy('order', 'asc').get().then(res =>{
9         this.setData({ boxlist: res.data })
10      })
```

```
11    },
12
13    enterapplication: function (event) {
14      var id = event.currentTarget.dataset.id
15      switch (id) {
16        case '通讯录':
17          wx.navigateTo({ url: '../contact/contact' })
18          break;
19        case '添加人员':
20          wx.navigateTo({ url: '../addperson/addperson' })
21          break;
22        case '删除人员':
23          wx.navigateTo({ url: '../deleteperson/deleteperson'})
24          break;
25        default:
26          break;
27      }
28    },
29  })
```

图 8-2　通讯录项目主页显示效果

代码第 7～11 行加载数据库集合 appbox 中的数据。代码第 13～28 行实现点击不同的应用图标跳转到不同的页面,本项目中使用 switch 语句根据数据 name 字段跳转到不同页面。要实现如图 8-2 所示的效果,需要在数据库集合 appbox 中添加 3 条数据,数据如下:

```
1   {
2       "_id":"dd1b1cae-a79f-4fbc-b27a-cc0713315133",
3       "order":1,
4       "url":"cloud://test-0pmu0.7465-test-0pmu0-1300559272/tontxunlu.png",
5       "name":"通讯录"
6   }
7   {
8       "_id":"1e2bc1d4-b79b-4001-86de-617b71da2f9e",
9       "name":"添加人员",
10      "order":2,
11      "url":"cloud://test-0pmu0.7465-test-0pmu0-1300559272/tianjiatongxunlu.png"
12  }
13  {
14      "_id":"980ac960-277e-4740-a38e-0635ab5efd5e",
15      "name":"删除人员",
16      "order":3,
17      "url":"cloud://test-0pmu0.7465-test-0pmu0-1300559272/shanchutongxunlu.png"
18  }
```

order 字段用于显示在页面的顺序，url 为图标的地址，所有图标均来自 https://www.iconfont.cn，用户选取图标下载后上传到项目的云存储上，把上面的 url 改成云存储上的图片地址。

8.2 通讯录页面

通讯录页面是本项目的核心页面，ColorUI 给出了和手机通讯录类似的页面样式，用微信开发者工具打开 ColorUI 中的 demo 文件，在 tabBar 中选择"扩展"选项，然后单击"索引列表"图标，进入"索引"页面（代码见 pages/plugin/indexes），ColorUI 通讯录样式如图 8-3 所示。

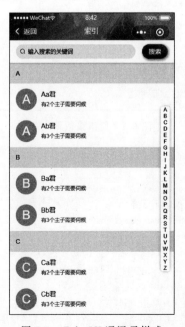

图 8-3　ColorUI 通讯录样式

为了便于看到通讯录的效果，本项目提前准备了 JSON 格式的数据（如何将 Excel 文件转换为 JSON 文件数据参见 3.2 节），导入数据库 contact 集合，导入后的结果如图 8-4 所示。其中 searchfield 字段是手工加进去的，是为了便于对人员信息进行模糊搜索，实现用户在输入框中输入姓名的中文、拼音或者拼音首字母便可以搜索人员，在添加人员页面将会讲解如何通过程序自动生成 searchfield 字段。

图 8-4　通讯录数据集合

本项目的通讯录页面 contact.wxml 的样式如下：

```
1   <cu-custom bgImage="/images/sylb2244.jpg" isBack="{{true}}">
2     <view slot="backText">返回</view>
3     <view slot="content">通讯录</view>
4   </cu-custom>
5
6   <view class="cu-bar bg-white search fixed" style="top:{{CustomBar}}px;">
7     <view class="search-form round">
8       <text class="cuIcon-search"></text>
9       <input type="text" placeholder="输入名字" confirm-type="search"
```

```
10          bindinput = "bindKeyInput"></input>
11      </view>
12      <view class = "action">
13          <button class = "cu-btn bg-gradual-green shadow-blur round" bindtap = 'search'>搜索</button>
14      </view>
15  </view>
16
17  <block wx:if = "{{issearch}}">
18      <view class = "cu-list menu-avatar no-padding">
19          <block wx:for = "{{contactCur}}" wx:key wx:for-index = "sub">
20              <view class = "cu-item" bindtap = "showModal" data-id = "{{sub}}" data-target = "bottomModal">
21                  <view class = "cu-avatar round lg">{{item.sex}}</view>
22                  <view class = "content">
23                      <view class = "text-grey">{{item.name}}
24                      </view>
25                      <view class = "text-gray text-sm">
26                          <text class = "text-abc">{{item.mobile}}</text>
27                      </view>
28                  </view>
29              </view>
30          </block>
31      </view>
32  </block>
33  <block wx:else>
34      <scroll-view scroll-y class = "indexes" scroll-into-view = "indexes-{{listCurID}}"
35  style = "height:calc(100vh - {{CustomBar}}px - 50px)" scroll-with-animation = "true"
36  enable-back-to-top = "true">
37      <block wx:for = "{{list}}" wx:key>
38          <block wx:if = "{{contactCur[index].length > 0}}">
39              <view class = "padding indexItem-{{list[index]}}" id = "indexes-{{list[index]}}"
40  data-index = "{{list[index]}}">{{list[index]}}</view>
41          </block>
42          <view class = "cu-list menu-avatar no-padding">
43              <block wx:for = "{{contactCur[index]}}" wx:key wx:for-index = "sub">
44                  <view class = "cu-item" bindtap = "showModal" id = "{{index}}" data-id = "{{sub}}"
45  data-target = "bottomModal">
46                      <view class = "cu-avatar round lg">{{item.sex}}</view>
47                      <view class = "content">
48                          <view class = "text-grey">{{item.name}}
49                          </view>
50                          <view class = "text-gray text-sm">
51                              <text class = "text-abc">{{item.mobile}}</text>
52                          </view>
53                      </view>
54                  </view>
55              </block>
56          </view>
57      </block>
58  </scroll-view>
```

```
59  <view class="indexBar" style="height:calc(100vh - {{CustomBar}}px - 50px)">
60    <view class="indexBar-box" bindtouchstart="tStart" bindtouchend="tEnd" catchtouchmove=
61  "tMove">
62      <view class="indexBar-item" wx:for="{{list}}" wx:key id="{{index}}" bindtouchstart="getCur"
63  bindtouchend="setCur">{{list[index]}}</view>
64    </view>
65  </view>
66  <!-- 选择显示 -->
67  <view hidden="{{hidden}}" class="indexToast">
68    {{listCur}}
69  </view>
70  </block>
71
72  <view class="cu-modal bottom-modal {{modalName=='bottomModal'?'show':''}}">
73    <view class="cu-dialog">
74      <view class="cu-bar bg-white">
75        <view class="action text-green">确定</view>
76        <view class="action text-blue" bindtap="hideModal">取消</view>
77      </view>
78      <view class="solid-bottom text-xl padding">
79        <block wx:if="{{!issearch}}">
80          <text class="text-black text-bold"
81  bindtap='makephone'>{{contactCur[selectid][selectsubid].mobile}}</text>
82        </block>
83        <block wx:else>
84          <text class="text-black text-bold"
85  bindtap='makephone'>{{contactCur[selectsubid].mobile}}</text>
86        </block>
87      </view>
88      <view class="solid-bottom text-xl padding">
89        <text class="text-black text-bold" bindtap='addcontact'>添加到通讯录</text>
90      </view>
91    </view>
92  </view>
```

代码第6~15行对应搜索输入框。由于搜索的结果不需要最右侧的索引,因此本页面代码第17~32行对应搜索结果显示的样式。代码第33~70行对应全部人员显示的样式,样式设计和ColorUI基本一致,包括用户性别、用户名字、用户手机号。代码第72~92行对应弹出框,允许用户把信息添加到手机通讯录,通讯录界面弹出框效果如图8-5所示。

通讯录页面为了显示索引效果,必须在contact.wxss文件中添加以下代码:

```
1  page {
2    padding-top: 100rpx;
3  }
4
5  .indexes {
6    position: relative;
7  }
```

图 8-5 通讯录界面弹出框效果

```
 8
 9   .indexBar {
10       position: fixed;
11       right: 0px;
12       bottom: 0px;
13       padding: 20rpx 20rpx 20rpx 60rpx;
14       display: flex;
15       align-items: center;
16   }
17
18   .indexBar .indexBar-box {
19       width: 40rpx;
20       height: auto;
21       background: #fff;
22       display: flex;
23       flex-direction: column;
24       box-shadow: 0 0 20rpx rgba(0, 0, 0, 0.1);
25       border-radius: 10rpx;
26   }
27
28   .indexBar-item {
29       flex: 1;
30       width: 40rpx;
31       height: 40rpx;
```

```
32      display: flex;
33      align-items: center;
34      justify-content: center;
35      font-size: 24rpx;
36      color: #888;
37    }
38
39    movable-view.indexBar-item {
40      width: 40rpx;
41      height: 40rpx;
42      z-index: 9;
43      position: relative;
44    }
45
46    movable-view.indexBar-item::before {
47      content: "";
48      display: block;
49      position: absolute;
50      left: 0;
51      top: 10rpx;
52      height: 20rpx;
53      width: 4rpx;
54      background-color: #f37b1d;
55    }
56
57    .indexToast {
58      position: fixed;
59      top: 0;
60      right: 80rpx;
61      bottom: 0;
62      background: rgba(0, 0, 0, 0.5);
63      width: 100rpx;
64      height: 100rpx;
65      border-radius: 10rpx;
66      margin: auto;
67      color: #fff;
68      line-height: 100rpx;
69      text-align: center;
70      font-size: 48rpx;
71    }
```

关于 WXSS 文件在这里不做过多说明，用户可自行查阅相关资料，本书侧重于项目的逻辑代码实现，对于样式可直接选用 ColorUI 中的样式。

相应的 contact.js 代码如下：

```
1  const app = getApp()
2  const db = wx.cloud.database()
3  Page({
4    data: {
```

```
5        StatusBar: app.globalData.StatusBar,
6        CustomBar: app.globalData.CustomBar,
7        hidden: true,
8        contactAll:[],
9        contactCur:[],
10       concatData:[],
11       issearch: false,
12       selectid: 0,
13       selectsubid: 0
14     },
15
16     onLoad: function(options) {
17       wx.showLoading({
18         title: '数据加载中...',
19       })
20       let list = []
21       var contactAll = []
22       for (let i = 0; i < 26; i++) {
23         list[i] = String.fromCharCode(65 + i)
24         contactAll[i] = new Array()
25       }
26       const MAX_LIMIT = 20
27       db.collection('contact').count().then(res =>{
28         const total = res.total
29         const batchTimes = Math.ceil(total / 20)
30         const tasks = []
31         for (let i = 0; i < batchTimes; i++) {
32           const promise = db.collection('contact').skip(i * MAX_LIMIT).limit(MAX_LIMIT).get();
33           tasks.push(promise)
34         }
35
36         Promise.all(tasks).then((result) =>{
37           let concatData = []
38           for (let i = 0; i < result.length; i++) {
39             concatData = concatData.concat(result[i].data)
40           }
41           for (let i = 0; i < concatData.length; i++) {
42             let index = concatData[i].searchfield[0].charCodeAt() - 97
43             contactAll[index].push(concatData[i])
44           }
45           this.setData({
46             list: list,
47             listCur: list[0],
48             contactAll: contactAll,
49             contactCur: contactAll,
50             concatData: concatData
51           })
52           wx.hideLoading()
53         })
```

```
54        })
55      },
56
57      bindKeyInput: function (event) {
58        var value = event.detail.value
59        value = value.toLowerCase()
60        var contactAll = this.data.contactAll
61        var concatData = this.data.concatData
62        var contactCur = []
63        if (event.detail.value == '') {
64          this.setData({
65            issearch: false,
66            contactCur: contactAll
67          })
68        }
69        else
70        {
71          for (let i = 0; i < concatData.length; i++) {
72            let index = concatData[i].searchfield.indexOf(value)
73            if(index! = -1)
74              contactCur.push(concatData[i])
75          }
76          console.log(contactCur)
77          this.setData({
78            issearch:true,
79            contactCur: contactCur
80          })
81        }
82      },
83
84      showModal(e) {
85        this.setData({
86          modalName: e.currentTarget.dataset.target,
87          selectid: e.currentTarget.id,
88          selectsubid: e.currentTarget.dataset.id
89        })
90      },
91      hideModal(e) {
92        this.setData({
93          modalName: null
94        })
95      },
96
97      makephone: function (event) {
98        var phoneNumber
99        if (! this.data.issearch) {
100         phoneNumber = this.data.contactCur[this.data.selectid][this.data.selectsubid].mobile
101       } else {
102         phoneNumber = this.data.contactCur[this.data.selectsubid].mobile
```

```
103      }
104      //console.log(phoneNumber)
105      wx.makePhoneCall({
106        phoneNumber: phoneNumber.toString()          //仅为示例,并非真实的电话号码
107      })
108    },
109
110    addcontact: function (event) {
111      var userinfo
112      if (!this.data.issearch) {
113        userinfo = this.data.contactCur[this.data.selectid][this.data.selectsubid]
114      } else {
115        userinfo = this.data.contactCur[this.data.selectsubid]
116      }
117      wx.addPhoneContact({
118        firstName: userinfo.name,
119        mobilePhoneNumber: userinfo.mobile.toString(),
120        success: function () {
121          console.log('添加成功')
122        }
123      })
124    },
125
126    onReady() {
127      let that = this;
128      wx.createSelectorQuery().select('.indexBar-box').boundingClientRect(function (res) {
129        that.setData({
130          boxTop: res.top
131        })
132      }).exec();
133      wx.createSelectorQuery().select('.indexes').boundingClientRect(function (res) {
134        that.setData({
135          barTop: res.top
136        })
137      }).exec()
138    },
139    //获取文字信息
140    getCur(e) {
141      this.setData({
142        hidden: false,
143        listCur: this.data.list[e.target.id],
144      })
145    },
146
147    setCur(e) {
148      this.setData({
149        hidden: true,
150        listCur: this.data.listCur
151      })
```

```javascript
152      },
153      //滑动选择Item
154      tMove(e) {
155        let y = e.touches[0].clientY,
156          offsettop = this.data.boxTop,
157          that = this;
158        //判断选择区域,只有在选择区域才会生效
159        if (y > offsettop) {
160          let num = parseInt((y - offsettop) / 20);
161          this.setData({
162            listCur: that.data.list[num]
163          })
164        };
165      },
166
167      //触发全部开始选择
168      tStart() {
169        this.setData({
170          hidden: false
171        })
172      },
173
174      //触发结束选择
175      tEnd() {
176        this.setData({
177          hidden: true,
178          listCurID: this.data.listCur
179        })
180      },
181      indexSelect(e) {
182        let that = this;
183        let barHeight = this.data.barHeight;
184        let list = this.data.list;
185        let scrollY = Math.ceil(list.length * e.detail.y / barHeight);
186        for (let i = 0; i < list.length; i++) {
187          if (scrollY < i + 1) {
188            that.setData({
189              listCur: list[i],
190              movableY: i * 20
191            })
192            return false
193          }
194        }
195      }
196    })
```

代码第 16～55 行对应页面加载事件 onLoad，需要在页面加载时读取数据库集合 contact 中所有人员的信息，因为该页面做了字母索引，所以需要把所有数据按照姓名首字母进行分类，直观的方法就是按照 26 个字母的顺序，依次查询数据库中姓名首字母符合要

求的查询操作,比如第一次查询数据库集合中姓名首字母为 a 的数据,第二次查询数据库集合中姓名首字母为 b 的数据。微信小程序端查询数据每次最多为 20 条记录,云函数端每次最多为 100 条记录,当数据量多的时候,每个字母对应的记录可能会超过 20 条,使得查询过程会比较复杂,而且当数据量少的时候,这种方式也需要执行 26 次数据库查询操作,显然增加了网络请求的消耗。因此这里首先把通讯录中的所有数据查询出来,然后每条数据按照姓名首字母进行分类,代码第 26~34 行把数据库集合 contact 中的数据都查询出来,因为微信小程序多次数据查询操作是异步操作,代码第 36 行采用 Promise.all() 方法来处理上面多个异步数据查询操作。随后代码第 37~40 行把上面的多次异步数据查询结果拼接在一起。代码第 41~44 行负责把所有的数据记录分成 26 类,放入 contactAll 数组中,contactAll 是一个包含了 26 个元素的数组,其中每个元素都包含了多个对象,AppData 中 contactAll 数据如图 8-6 所示。需要注意的是,在数据分类时,按照姓名首字母(searchfield 字段第一个字符)进行分类,通过 push 操作把记录保存到 contactAll 数组中,这里 contactAll 数组是一个二重数组,因此在代码第 24 行需要把 contactAll 初始化为二重数组。

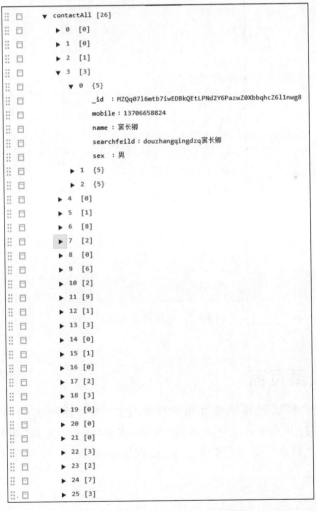

图 8-6　AppData 中 contactAll 数据

该页面另一个很重要的功能是查询功能，代码第 57~82 行是查询操作，需要进行多次数据查询，如果每次查询都去读取数据库，显然会降低该页面的执行效率，而且在 onLoad 事件中已读取所有人员的数据，因此查询功能中使用了 indexOf 操作（indexOf( )方法可返回某个指定的字符串值在字符串中首次出现的位置）直接查询 contactData 中是否有数据，这样就不用读取云数据库中的数据，提高了效率，当搜索框中没有输入时，页面显示所有人员信息。代码第 72 行从数组 contactData 中逐个搜索每个对象的 searchfield 中是否含有输入框中的字符串，如果含有输入框中的字符串，代码第 74 行把该元素加入 contactCur 数组中。代码第 84~95 行对应显示和隐藏模态对话框。代码第 97~108 行对应点击图 8-5 中电话号码实现拨打电话。代码第 110~124 行对应单击图 8-5 中"添加到通讯录"实现添加到手机通讯录的功能。代码第 126~196 行对应 ColorUI 中实现索引效果的功能。通讯录页面效果如图 8-7 所示。

图 8-7　通讯录页面效果

8.3　删除人员页面

删除人员页面的样式和通讯录页面的样式类似，该页面使用了列表左滑实现删除。用微信开发者工具打开 ColorUI 中的 demo 文件，在 tabBar 中选择"组件"选项，然后单击"列表"选项，进入"列表"页面（代码见 pages/component/list），左滑删除样式如图 8-8 所示。

图 8-8　左滑删除样式

删除页面 deleteperson.wxml 代码如下：

```
1   <cu-custom bgImage = "/images/sylb2244.jpg" isBack = "{{true}}">
2       <view slot = "backText">返回</view>
3       <view slot = "content">删除人员</view>
4   </cu-custom>
5
6   <view class = "cu-bar bg-white search fixed" style = "top:{{CustomBar}}px;">
7       <view class = "search-form round">
8           <text class = "cuIcon-search"></text>
9           <input type = "text" placeholder = "输入名字" confirm-type = "search"
10  bindinput = "bindKeyInput"></input>
11      </view>
12      <view class = "action">
13          <button class = "cu-btn bg-gradual-green shadow-blur round" bindtap = 'search'>搜索</button>
14      </view>
15  </view>
16
17  <view class = "cu-list menu-avatar no-padding" style = "padding-top: 100rpx;">
18      <view class = "cu-item {{modalName == 'move-box-' + index?'move-cur':''}}" wx:for = "{{contactCur}}"
19  wx:key bindtouchstart = "ListTouchStart" bindtouchmove = "ListTouchMove"
20  bindtouchend = "ListTouchEnd" data-target = "move-box-{{index}}">
21          <view class = "cu-avatar round lg">{{item.sex}}</view>
22          <view class = "content">
23              <view class = "text-grey">{{item.name}}
```

```
24          </view>
25          <view class = "text - gray text - sm">
26            <text class = "text - abc">{{item.mobile}}</text>
27          </view>
28        </view>
29        <view class = "move">
30          <view class = "bg - grey">置顶</view>
31          <view class = "bg - red" bindtap = "deleteperson" data - id = "{{index}}">删除</view>
32        </view>
33      </view>
34  </view>
```

代码第 6~15 行对应搜索输入框，在该页面中搜索输入框的功能并没有实现。代码第 17~34 行显示通讯录数据，加入了列表左滑出现"置顶"和"删除"操作，本页面仅实现了删除功能，通讯录删除人员页面如图 8-9 所示，用户在列表中左滑实现该记录删除操作。

图 8-9　通讯录删除人员页面

相应的 deleteperson.js 代码如下：

```
1  const app = getApp();
2  const db = wx.cloud.database()
3  Page({
4    data: {
5      StatusBar: app.globalData.StatusBar,
6      CustomBar: app.globalData.CustomBar,
7      contactCur: [],
```

```
8        page: 0,
9      },
10
11     onLoad: function (options) {
12       wx.showLoading({
13         title: '数据加载中...',
14       })
15       db.collection('contact').get().then(res =>{
16         this.setData({ contactCur: res.data })
17       })
18       wx.hideLoading()
19     },
20
21     bindKeyInput: function (event) {
22     },
23
24     // ListTouch 触摸开始
25     ListTouchStart(e) {
26       this.setData({
27         ListTouchStart: e.touches[0].pageX
28       })
29     },
30
31     // ListTouch 计算方向
32     ListTouchMove(e) {
33       this.setData({
34         ListTouchDirection: e.touches[0].pageX - this.data.ListTouchStart > 0 ? 'right' : 'left'
35       })
36     },
37
38     // ListTouch 计算滚动
39     ListTouchEnd(e) {
40       if (this.data.ListTouchDirection == 'left') {
41         this.setData({
42           modalName: e.currentTarget.dataset.target
43         })
44       } else {
45         this.setData({
46           modalName: null
47         })
48       }
49       this.setData({
50         ListTouchDirection: null
51       })
52     },
53
54     onReachBottom: function () {
55       var page = this.data.page + 20
56       db.collection('contact').skip(page).get().then(res =>{
```

```
57        console.log(res.data)
58        var contactCur = this.data.contactCur
59        contactCur = contactCur.concat(res.data)
60        this.setData({ contactCur: contactCur, page: page })
61      })
62    },
63
64    deleteperson: function (event) {
65      var id = event.currentTarget.dataset.id
66      var contactCur = this.data.contactCur
67      var _id = contactCur[id]._id
68      console.log(_id)
69      contactCur.splice(id, 1)
70      this.setData({ contactCur: contactCur })
71
72      wx.cloud.callFunction({
73        name: 'deleteperson',
74        data: {
75          _id: _id
76        },
77        success: function (res) {
78          wx.showToast({
79            title: '删除成功...',
80            icon: 'none',
81            duration: 1500
82          })
83        },
84        fail: console.error
85      })
86    },
87  })
```

代码第 11~19 行对应页面加载事件，删除人员页面没有使用索引列表，因此不需要像通讯录页面那样一次把所有人员的信息都加载进来，删除人员页面仅需要加载前 20 条记录，当用户浏览完这 20 条记录时，会触发 onReachBottom 事件。在代码第 54~62 行 onReachBottom 中，通过使用 db.collection('contact').skip(page)查询指令跳过 page 条记录，随后通过 contact 命令，把查询的结果合并到现有的人员信息 contactCur 中。在本页面中并没有使用输入框搜索功能，代码第 24~52 行对应 ColorUI 中实现左滑出现"置顶"和"删除"操作的代码，代码第 64~86 行实现单击"删除"按钮，删除相应的记录。删除操作进行 2 步操作：第 1 步需要在现有人员信息 contactCur 中，删除选中的记录；第 2 步调用云函数 deleteperson 删除云数据库 contact 集合中选中的记录。需要说明的是为什么使用云函数来删除数据库中的记录，因为数据库集合 contact 的权限设置为"所有用户可读，仅创建者可读写"，而本项目的人员信息大部分是通过数据库导入的方式导入的，每条记录中并没有_openid 字段，也就是说即使这些记录是管理员本人导入的，也没办法通过微信小程序端来执行删除操作，一种方式是在数据导入之前每条记录增加_openid 字段。但是很多时候管理员并不是一个，也就是说执行删除操作的人员并不是数据创建者时，就必须使用云函数。使用的云函数 deleteperson 代码如下：

```
1   const cloud = require('wx-server-sdk')
2   cloud.init()
3   const db = cloud.database()
4   exports.main = async (event, context) =>{
5     var _id = event._id
6     try {
7       return await db.collection('contact').doc(_id).remove()
8     } catch (e) {
9       console.error(e)
10    }
11  }
```

8.4 添加人员页面

添加人员页面 addperson.wxml 代码如下：

```
1   <cu-custom bgImage = "/images/sylb2244.jpg" isBack = "{{true}}">
2     <view slot = "backText">返回</view>
3     <view slot = "content">添加人员</view>
4   </cu-custom>
5
6   <form>
7     <view class = "cu-form-group">
8       <view class = "title">
9         <text style = "color:red">*</text>姓名:</view>
10        <input bindinput = "updateValue" data-name = 'name' placeholder = "请输入姓名"></input>
11    </view>
12    <view class = "cu-form-group">
13      <view class = "title">
14        <text style = "color:red">*</text>手机:</view>
15        <input bindinput = "updateValue" data-name = 'mobile' placeholder = "请输入手机号"></input>
16    </view>
17    <view class = "cu-form-group">
18      <view class = "title">
19        <text style = "color:red">*</text>性别:</view>
20        <picker bindchange = "PickerChange" value = "{{index}}" range = "{{picker}}">
21          <view class = "picker">
22            {{picker[index]}}
23          </view>
24        </picker>
25    </view>
26  </form>
27
28  <view class = "flex padding justify-center">
29    <button class = "cu-btn lg bg-green" bindtap = "submitform">提交</button>
30  </view>
```

添加人员页面的显示效果如图 8-10 所示。

图 8-10　添加人员页面的显示效果

相应的 addperson.js 代码如下：

```
1   const db = wx.cloud.database()
2   Page({
3     data:{
4       index:0,
5       picker:['男','女'],
6       personInfo:{ name:'', mobile:'',sex:'男'}
7     },
8     PickerChange(e){
9       let index = e.detail.value
10      let personInfo = this.data.personInfo
11      personInfo.sex = index==1?'女':'男'
12      this.setData({
13        index:index,personInfo:personInfo
14      })
15    },
16    updateValue:function(event){
17      let name = event.currentTarget.dataset.name;
18      let personInfo = this.data.personInfo
19      personInfo[name] = event.detail.value
20      this.setData({
21        personInfo:personInfo
22      })
```

```
23    },
24
25    submitform: function (event) {
26      var mobile = this.data.personInfo.mobile;
27      var name = this.data.personInfo.name;
28      var sex = this.data.personInfo.sex;
29      var mobilevalid = /^(((13[0-9]{1})|(15[0-9]{1})|(18[0-9]{1})|(17[0-9]{1}))+\d{8})$/;
30      if (name == '') {
31        wx.showToast({
32          title: '请输入用户名',
33          icon: 'succes',
34          duration: 1000,
35          mask: true
36        })
37        return false
38      } else if (mobile == '') {
39        wx.showToast({
40          title: '手机号不能为空',
41        })
42        return false
43      }
44      else if (mobile.length != 11) {
45        wx.showToast({
46          title: '手机号长度有误!',
47          icon: 'success',
48          duration: 1500
49        })
50        return false;
51      }
52      if (!mobilevalid.test(mobile)) {
53        wx.showToast({
54          title: '手机号有误!',
55          icon: 'success',
56          duration: 1500
57        })
58        return false;
59      }
60      wx.cloud.callFunction({
61        name: 'translate',
62        data: {
63          text: name
64        },
65        success: function (res) {
66          console.log(res.result.searchfield)
67          db.collection('contact').add({
68            data: {
69              name: name,
70              mobile:mobile,
71              sex: sex,
```

```
72              searchfield: res.result.searchfield
73            }
74          })
75          .then(res =>{
76            wx.showToast({
77              title: '添加成功...',
78              icon: 'none',
79              duration: 1500
80            })
81          })
82      }
83    })
84  }
85 })
```

代码第 8～15 行获取性别选择器的值。代码第 16～23 行获取姓名、手机输入框的值。本页面的核心代码为 submitform 事件，首先项目对输入的用户名和密码做了校验，代码第 31～43 行分别判断姓名和手机号是否为空，代码第 44～51 行判断手机号码是否为 11 位，代码第 52～59 行通过正则表达式验证手机号码是否合法。接下来一个关键的步骤是获取姓名的拼音，从而获得 searchfield 字段，通过云函数 translate 把姓名转换为本项目所需要的 searchfield 字段，然后通过 db.collection('contact').add 操作把数据提交到云数据库 contact 集合中。

通过使用音译(transliteration)模块把中文转换为拼音，该模块支持中文转拼音或生成 slug，见 https://github.com/dzcpy/transliteration。

安装音译(transliteration)模块：

```
1 npm install transliteration -- save
```

安装完音译(transliteration)模块后，在云函数 package.json 中可以看到安装了相应的依赖库：

```
1 "dependencies": {
2   "transliteration": "^2.1.7",
3   "wx-server-sdk": "^1.5.3"
4 }
```

本页面的 translate 云函数代码如下：

```
1 const cloud = require('wx-server-sdk')
2 var transliteration = require('transliteration');
3 cloud.init()
4 exports.main = async (event, context) =>{
5   var text = event.text
6   var pinyin = transliteration.slugify(text, { lowercase: true, separator: '_' })
7   var pinyin_arr = pinyin.split("_");
```

```
8    var searchfield = pinyin.replace(/_/g,"")
9    for(i=0;i<pinyin_arr.length;i++)
10   {
11     searchfield = searchfield + pinyin_arr[i][0]
12   }
13   searchfield = searchfield + text
14   return {
15     searchfield
16   }
17 }
```

代码第 6 行把中文 text 转换为小写的拼音，每个拼音之间用_连接，比如"transliteration.slugify('你好,世界',{lowercase:false,separator:'_'});//ni_hao_shi_jie"。代码根据连接符把每个汉字的拼音都存储在数组中，云函数根据输入的中文名字，获取 searchfield 字段，searchfield 字段的组成为：拼音全称＋每个汉字首字母＋汉字名称；代码中第 8 行获得拼音全称，即把音译(transliteration)模块输出值中的连接符去掉，代码第 9～12 行提取每个拼音的首字母，代码 13 行把汉字名称加到 searchfield 字段中。比如云函数的输入汉字姓名为蹇向秋，那么最后 searchfield 字段的值为"jianxiangqiujxq 蹇向秋"。

第 9 章

报修微信小程序

报修微信小程序包含 6 个页面：

```
1   "pages": [
2     "pages/login/login",                    //登录页面
3     "pages/index/index",                    //项目主页
4     "pages/repairtab/repairtab",            //用户报修页面
5     "pages/repairtabadmin/repairtabadmin",  //管理员管理界面
6     "pages/orderform/orderform",            //填写报修单页面
7     "pages/formdetail/formdetail",          //报修单页面
8   ],
```

其中包含 2 个数据库集合 Verify 和 order，权限均为"所有用户可读，仅创建者可读写"。其中 Verify 集合存储用户登录记录，order 集合存储报修单记录。

9.1 腾讯云短信平台

因为要使用短信验证码登录，本项目使用的是腾讯云短信平台（https://cloud.tencent.com/document/product/382），个人用户可以登录腾讯云控制台（https://cloud.tencent.com/），开通腾讯云短信服务，一般情况下腾讯云平台会赠送短信套餐包给用户，国内短信固定套餐包具体价格如表 9-1 所示。

表 9-1 国内短信固定套餐包具体价格

用户认证类型	短信条数/万条	套餐售价/元	套餐内单价/元每条	支持的短信类型
企业认证用户	1	470	0.045	国内验证码短信
	10	4 200	0.042	国内通知类短信
	50	20 500	0.041	国内营销类短信

续表

用户认证类型	短信条数/万条	套餐售价/元	套餐内单价/元每条	支持的短信类型
企业认证用户	100	40 000	0.040	国内验证码短信 国内通知类短信 国内营销类短信
	300	117 000	0.039	
个人认证用户	1	470	0.045	国内验证码短信 国内通知类短信
	10	4 200	0.042	
	50	20 500	0.041	
	100	40 000	0.040	
	300	117 000	0.039	

用户想了解腾讯云短信具体信息请自行查看腾讯云短信平台文档。

在微信小程序中使用腾讯云短信需要准备如下信息：

1. 获取 SDK AppID 和 App Key

云短信应用 SDK AppID 和 App Key 可在短信控制台的应用信息中获取。如果尚未添加应用，则登录短信控制台添加应用。腾讯云短信应用页面如图9-1所示。

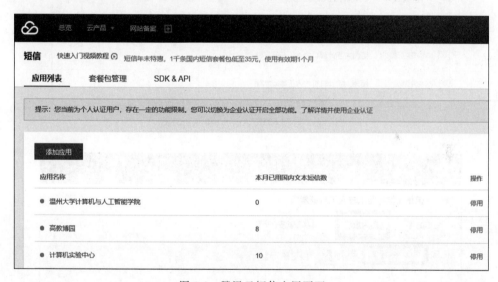

图9-1 腾讯云短信应用页面

添加成功应用后，单击页面中的应用名称，进入应用后选择"应用配置"→"基础配置"选项，查看应用的 SDK AppID 和 App Key，如图9-2所示。项目开发中发送短信需要使用 SDK AppID 和 App Key。

2. 申请签名并确认审核通过

一个完整的短信由短信签名和短信正文内容两部分组成，短信签名需申请和审核，签名可在短信控制台的相应服务模块"短信内容配置"中进行申请，如图9-3所示，国内短信由签名＋正文组成，签名符号为【】(注：全角)，发送短信内容时必须带签名。创建成功应用后，进入应用后选择"国内短信"→"短信内容配置"→"短信签名"选项，单击"创建签名"按钮。

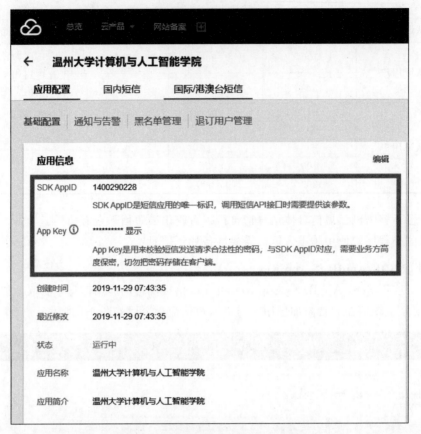

图 9-2 查询 SDK AppID 和 App Key

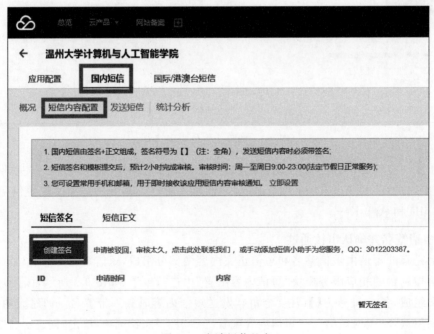

图 9-3 申请短信签名

用户填写创建短信签名，如图 9-4 所示，在这里签名类型选择"公众号或微信小程序"。需要说明的是，这里需要上传证明文件，证明文件为微信公众平台管理界面截图，也就是说微信小程序必须上线运行。用户可以先上线一个简化版本，上线后登录微信公众平台官方网站 https://mp.weixin.qq.com，进入后在左侧选择"设置"选项，选择基本设置子页面，上传页面截图作为证明文件，如果在申请过程中遇到困难，也可以通过 QQ 联系短信平台联系人。

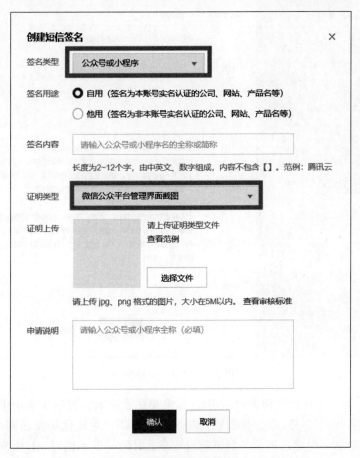

图 9-4　创建短信签名

3. 申请模板并确认审核通过

短信正文内容模板需申请和审核，模板可在短信控制台的相应服务模块"内容配置"中进行申请，如图 9-5 所示，进入应用后选择"国内短信"→"短信内容配置"→"短信正文"选项，单击"创建正文模板"按钮，打开创建短信正文模板页。

短信模板示例：{1}为您的登录验证码，请于{2}分钟内填写。如非本人操作，请忽略本短信(其中{数字}为可自定义的内容，须从 1 开始连续编号，如{1}、{2}等)。本项目需要申请 3 个短信模板，申请的短信正文如图 9-6 所示，其中短信 ID 为 386540 对应用户登录微信小程序时发送的短信验证码，短信 ID 为 386955 对应管理员对维修单进行派修后发给用户的短信模板，短信 ID 为 386954 对应派修单完成后发给用户的短信模板。用户在编写发送

图 9-5　申请短信模板

短信程序时,需要使用自己申请的短信 ID。这里要注意腾讯云短信正文中单个变量字数个人账号最多支持 12 个字符,而企业账号无限制。个人账户对短信的发送频率也做了限制:对同一个手机号,30s 内发送短信条数不超过 1 条;对同一个手机号,1h 内发送短信条数不超过 5 条;对同一个手机号,1 个自然日内发送短信条数不超过 10 条。当然腾讯云也给出了频率限制白名单:在白名单的手机号码发送短信不受频率限制策略的影响,仍可正常发送。最多可以添加 300 个白名单号码,该设置仅对国内短信和国际及港澳台地区短信生效。

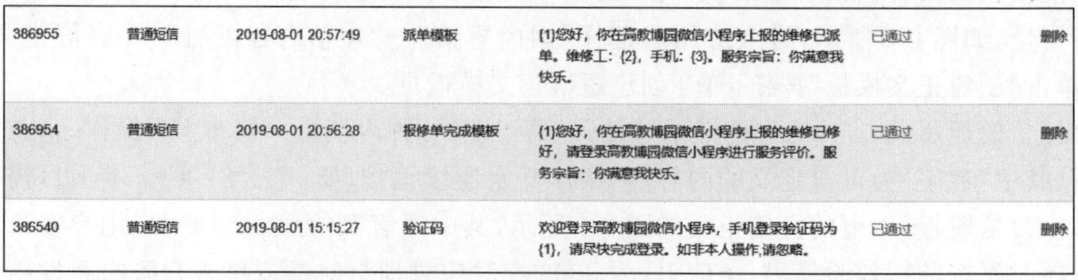

图 9-6　申请的短信正文

9.2 登录页面

本项目登录页面采用手机验证码登录方式,登录页面 login.wxml 代码如下:

```
1   <cu-custom bgColor = "bg-gradual-pink" isBack = "{{false}}">
2     <view slot = "content">登录页面</view>
3   </cu-custom>
4
5   <image style = "width:100%" mode = 'scaleToFill' src = "/images/bg.jpg"></image>
6   <view class = "cu-form-group">
7     <view class = "title">
8       <text style = "color:red">*</text>手机号:</view>
9     <input placeholder = "请输入手机号码" type = "number" maxlength = "11"
10  bindinput = "hindleInputPhone"></input>
11  </view>
12  <view class = "cu-form-group">
13    <view class = "title">
14      <text style = "color:red">*</text>密码:</view>
15    <input placeholder = "验证码" type = "number" maxlength = "6" bindinput = "handleInputVerify"
16  value = "{{verify}}" disabled = "{{! verifyStatus}}"></input>
17    <button class = "cu-btn bg-green shadow" disabled type = "" disabled = "{{! phoneStatus}}"
18  bindtap = "hindleSendVerify">{{verifyButtonText}}</button>
19  </view>
20  <view style = 'display:flex;justify-content:center;'>
21    <button class = "cu-btn lg bg-pink" bindgetuserinfo = 'login' disabled type = "" disabled =
22  "{{! loginStatus}}" open-type = "getUserInfo">登录</button>
23  </view>
24
25  <view style = "position:fixed;width:100%;bottom:0;display:flex;justify-content:center;">
26    <text class = "text-df text-darkGray">Copyright ? 温州大学 Wenzhou University All
27  Rights Reserved.</text>
28  </view>
```

登录页面的样式相对比较简单,只有两个输入框,要求用户输入手机号和密码,代码第 5 行放了一张图片,为了便于读者看到效果,图片放在了本地目录中,实际开发中读者应把图片放到云存储中,登录页面的效果如图 9-7 所示。

用户输入手机号后,单击"验证码"按钮,这时用户手机上就会收到一条验证码短信,如图 9-8 所示。用户把收到的短信验证码输入登录页面,然后可单击"登录"按钮进行登录。

相应的 login.js 代码如下:

```
1   Page({
2     data:{
3       phoneStatus:false,
4       verifyStatus:false,
5       loginStatus:false,
6       verifyButtonText:'验证码',
```

图 9-7 报修微信小程序登录页面的效果

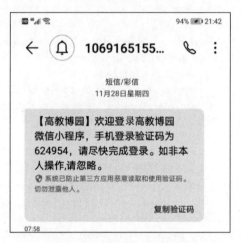

图 9-8 用户验证码短信

```
 7        phone: '',
 8        verify: '',
 9        second: 60,
10        verifyCount: 0,
11        canIUse: wx.canIUse('button.open-type.getUserInfo')
12      },
13
14      hindleInputPhone: function (event) {
15        let phone = event.detail.value
16        if (phone.length === 11) {
17          if (this.checkPhone(phone)) {
18            wx.hideKeyboard()
19            this.setData({
20              phone: phone
21            })
22            this.showVerifyButton()
23          } else {
24            wx.showToast({
25              title: '手机号格式异常',
26              icon: 'none',
27              duration: 1500
28            })
29          }
30        } else {
```

```
31        this.setData({
32          phone: phone,
33          phoneStatus: false
34        })
35      }
36    },
37
38    checkPhone: function (phone) {
39      let str = /^1\d{10}$/
40      if (str.test(phone)) {
41        return true
42      } else {
43        return false
44      }
45    },
46
47    showVerifyButton: function () {
48      if (this.data.verifyButtonText === '验证码') {
49        this.setData({
50          phoneStatus: true
51        })
52      }
53    },
54
55    hindleSendVerify: function () {
56      //console.log('获取验证码')
57      let self = this
58      self.setData({
59        phoneStatus: false
60      })
61      wx.cloud.callFunction({
62        name: 'sendsms',
63        data: {
64          phone: this.data.phone,
65          templateId: 386540
66        },
67        success: res =>{
68          console.log(res)
69          if (res.result) {
70            wx.showToast({
71              title: res.result.msg,
72              icon: 'none',
73              duration: 1500,
74            })
75            if (res.result.code === 0) {
76              this.setData({
77                phoneStatus: false,
78                verifyStatus: true,
79                verifyCount: this.verifyCount + 1,
```

```js
              verify: ''
            })
            this.timer()
          } else {
            self.setData({
              phoneStatus: true
            })
          }
        } else {
          wx.showToast({
            title: '验证码发送失败',
            icon: 'none',
            duration: 1500
          })
          self.setData({
            phoneStatus: true
          })
        }
      }
    })
  },

  timer: function () {
    let promise = new Promise((resolve, reject) =>{
      let setTimer = setInterval(
        () =>{
          this.setData({
            verifyButtonText: `${this.data.second - 1}s`,
            second: this.data.second - 1,
          })
          if (this.data.second <= 0) {
            if (this.checkPhone(this.data.phone)) {
              this.setData({
                second: 60,
                verifyButtonText: '验证码',
                phoneStatus: true
              })
            } else {
              this.setData({
                second: 60,
                verifyButtonText: '验证码',
                phoneStatus: false,
                verifyStatus: false
              })
            }
            resolve(setTimer)
          }
        }, 1000
      )
```

```
129        })
130        promise.then((setTimer) =>{
131          clearInterval(setTimer)
132        })
133    },
134
135    handleInputVerify: function (e) {
136      let verify = e.detail.value
137      if (verify.length === 6) {
138        this.setData({
139          verify: verify,
140          loginStatus: true
141        })
142        wx.hideKeyboard()
143      }
144    },
145
146    login: function (event) {
147      if (event.detail.userInfo) {
148        let self = this
149        self.setData({
150          loginStatus: false
151        })
152        wx.setStorageSync('userInfo', event.detail.userInfo);
153        let phone = this.data.phone
154        let verify = this.data.verify
155        wx.cloud.callFunction({
156          name: 'checksms',
157          data: {
158            phone,
159            verify
160          },
161          success: res =>{
162            console.log(res)
163            wx.showToast({
164              title: res.result.msg,
165              icon: 'none',
166              duration: 1500
167            })
168            if (res.result.code === 0) {
169              wx.setStorageSync('phone', phone);
170              wx.navigateTo({
171                url: '../index/index'
172              })
173            } else {
174              self.setData({
175                loginStatus: true
176              })
177            }
```

```
178            }
179         })
180      }
181   },
182 })
```

代码第 14~36 行对应手机号输入框,当手机号码长度达到 11 位时,验证手机号是否合法。代码第 38~45 行 checkPhone 函数采用正则表达式验证手机号是否合法,如果合法则设置 phoneStatus 为 true,这时"验证码"按钮变成可使用状态。代码第 55~100 行对应单击"验证码"按钮事件,其中代码第 61~67 行调用云函数 sendsms 发送短信验证码,短信模板 ID 为 386540,对应在腾讯云平台申请的短信模板,如果短信验证码发送成功,代码第 102~133 行 timer 函数设置定时器 setInterval 为 60s,在 60s 内"验证码"按钮为禁用状态,防止用户多次单击"验证码"按钮,代码第 135~144 行对应输入验证码,当输入验证码长度为 6 位时,"登录"按钮从禁用状态转变为可使用状态,代码第 146~181 行对应单击"登录"按钮事件,代码第 155~160 行调用云函数 checksms 验证用户输入的手机号和验证码是否正确,如果正确则跳转到项目主页 index。需要说明的是"登录"按钮 open-type ="getUserInfo",因此在使用"登录"按钮时,先检测 event.detail.userInfo 是否获得微信授权,如果已经获得微信授权,则继续判断手机号和验证码是否正确。

在登录页面中调用 sendsms 云函数发送手机验证码和 checksms 云函数验证手机号验证码是否正确,sendsms 云函数的代码如下:

```
1  const cloud = require('wx-server-sdk')
2  const QcloudSms = require("qcloudsms_js")
3  const appid = ***********  //替换成用户申请的云短信 AppID 以及 App Key
4  const appkey = "********************"//替换成用户申请的云短信 AppID 以及 App Key
5  const smsSign = ""
6  const uinSendNum = 3
7  cloud.init()
8  const db = cloud.database()
9
10 exports.main = async (event, context) => new Promise((resolve, reject) =>{
11   const wxContext = cloud.getWXContext()
12   var {
13     phone,
14     templateId
15   } = event
16   console.log(phone, templateId)
17   if (templateId == 386540) {        //替换成用户申请的模板 ID
18     db.collection('Verify').where({
19       phone: phone
20     }).orderBy('sendtime', 'desc').limit(uinSendNum).get().then(res =>{
21       if (res.data.length === 3 && new Date(res.data[2].sendtime).toDateString() === new Date().toDateString()) {
22
23         console.log("uin 超限:")
24         resolve({
```

```
25            code: 1013,
26            msg: '您频繁发送手机验证码,请稍后再试'
27          })
28        } else {
29          let qcloudsms = QcloudSms(appid, appkey)
30          let ssender = qcloudsms.SmsSingleSender()
31          let verify = ('000000' + Math.floor(Math.random() * 999999)).slice(-6)
32          let params = [verify]
33          ssender.sendWithParam(86, event.phone, templateId,
34            params, smsSign, "", "", (err, res, resData) =>{
35              if (err) {
36                console.log("err:", err);
37                reject({
38                  err
39                })
40              } else {
41                if (resData.result === 0) {
42                  var {
43                    result,
44                    errmsg,
45                    fee,
46                    sid
47                  } = resData
48                } else {
49                  var {
50                    result,
51                    errmsg
52                  } = resData
53                }
54                db.collection('Verify').add({
55                  data: {
56                    phone: event.phone,
57                    verify: verify,
58                    activetime: 5,
59                    sendtime: new Date().getTime(),
60                    senddate: new Date(),
61                    fee: fee ? fee : '',
62                    sid: sid ? sid : '',
63                    sendResult: result,
64                    errmsg: errmsg,
65                    recvtime: '',
66                    recvmsg: '',
67                    recvresult: '',
68                    inpuTime: '',
69                    inputVerify: '',
70                    inputCount: parseInt(0),
71                    verifyStatus: false
72                  }
73                }).then(res =>{
```

```javascript
74            resolve({
75              code: result,
76              msg: result === 0 ? '验证码发送成功' : errmsg
77            })
78          })
79        }
80      }
81    )
82   }
83  })
84 }
85
86 if (templateId == 386955) {
87   console.log(templateId, phone)
88   let qcloudsms = QcloudSms(appid, appkey)
89   let ssender = qcloudsms.SmsSingleSender()
90   let name = event.name
91   let staff = event.staff
92   let staffphone = event.staffphone
93   let params = [name, staff, staffphone]
94   ssender.sendWithParam(86, event.phone, templateId,
95     params, smsSign, "", "", (err, res, resData) =>{
96       if (err) {
97         console.log("err: ", err);
98         reject({
99           err
100        })
101      } else {
102        if (resData.result === 0) {
103          var {
104            result,
105            errmsg,
106            fee,
107            sid
108          } = resData
109        } else {
110          var {
111            result,
112            errmsg
113          } = resData
114        }
115      }
116    }
117  )
118 }
119
120 if (templateId == 386954) {
121   console.log(templateId, phone)
122   let qcloudsms = QcloudSms(appid, appkey)
```

```
123        let ssender = qcloudsms.SmsSingleSender()
124        let name = event.name
125        let params = [name]
126        ssender.sendWithParam(86, event.phone, templateId,
127          params, smsSign, "", "", (err, res, resData) =>{
128            if (err) {
129              console.log("err:", err);
130              reject({
131                err
132              })
133            } else {
134              if (resData.result === 0) {
135                var {
136                  result,
137                  errmsg,
138                  fee,
139                  sid
140                } = resData
141              } else {
142                var {
143                  result,
144                  errmsg
145                } = resData
146              }
147            }
148          }
149        )
150      }
151  })
```

用户需要在代码第 3、4 行填写腾讯云平台上申请的 AppID 和 App Key。sendsms 为发送短信云函数，包含了可以发送短信模板 ID 为 386540、386955、386954 共 3 种短信模板，因为发送短信需要给出短信模板 ID，所以 sendsms 并没有采用 TcbRouter 的方式写云函数，这里采用 if 或者 switch 语句根据 templatedId 来区分调用不同的短信模板。代码第 17～84 行对应发送短信验证码，其中代码第 21 行对短信的发送频率做了限制，获取数据库 Verify 中本人发送的最近 3 条验证码，如果当天发送验证码超过 3 条，则会提示："您频繁发送手机验证码，请稍后再试"。开发者在测试的时候先去掉这个限制，等登录功能正常后，再把这个限制加上去。代码第 29～81 行调用腾讯云发送短信 API 函数 ssender.sendWithParam 发送手机短信，代码第 31 行随机产生 6 位数字作为手机验证码，手机验证码发送成功后，代码第 54～78 行把相应的记录写入数据库 Verify 集合。代码第 86～118 行对应发送模板 ID 为 386955 的短信，代码第 120～150 行对应发送模板 ID 为 386954 的短信。这里需要说明的是，sendsms 调用需要依赖于 qcloudsms_js 模板，因此需要在 sendsms 云函数的 package.json 依赖库中加入 qcloudsms_js：

```
1   "dependencies": {
2     "wx-server-sdk": "latest",
3     "qcloudsms_js": "^0.1.1"           //加入 qcloudsms_js 依赖库
4   }
```

验证码发送成功后,进入云开发控制台,数据库集合 Verify 中就会增加一条短信验证码的数据,如图 9-9 所示。

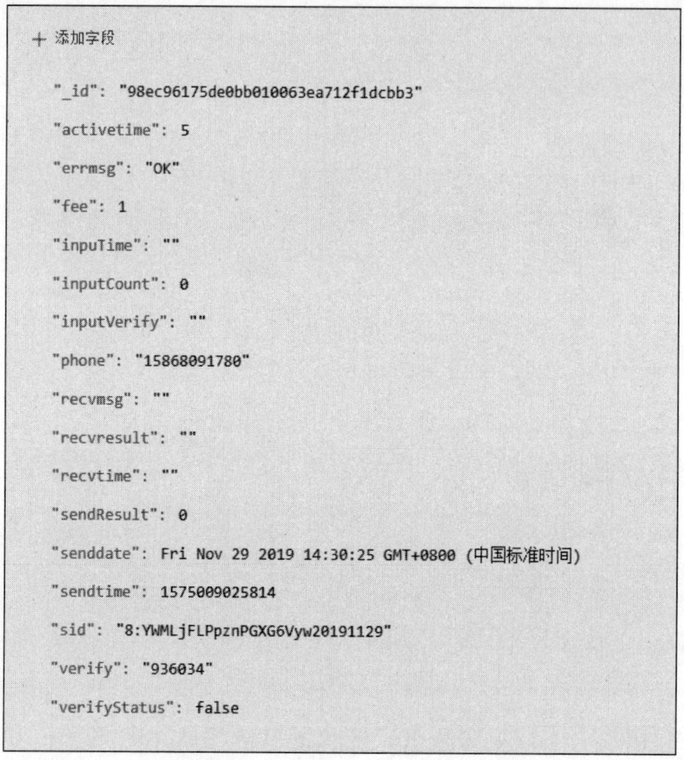

```
+ 添加字段

"_id": "98ec96175de0bb010063ea712f1dcbb3"
"activetime": 5
"errmsg": "OK"
"fee": 1
"inpuTime": ""
"inputCount": 0
"inputVerify": ""
"phone": "15868091780"
"recvmsg": ""
"recvresult": ""
"recvtime": ""
"sendResult": 0
"senddate": Fri Nov 29 2019 14:30:25 GMT+0800 (中国标准时间)
"sendtime": 1575009025814
"sid": "8:YwMLjFLPpznPGXG6Vyw20191129"
"verify": "936034"
"verifyStatus": false
```

图 9-9 数据库短信验证码的数据

云函数 checksms 验证用户输入的手机号和验证码是否正确,云函数 checksms 的代码如下:

```
1   const cloud = require('wx-server-sdk')
2   cloud.init()
3   const db = cloud.database()
4   const _ = db.command
5   exports.main = async (event, context) =>{
6     const wxContext = cloud.getWXContext()
7     var { phone, verify } = event
8     console.log(phone, verify)
9     if (phone && verify) {
10      let code = 0
11      let msg = '验证成功'
12      let time = new Date().getTime()
```

```
13      let data = await db.collection('Verify').where({
14        phone: phone + ""
15      }).orderBy('sendtime', 'desc').limit(1).get()
16      console.log("data:", data)
17      let updataData = {
18        inpuTime: time,
19        inputCount: parseInt(data.data[0].inputCount) + 1,
20        inputVerify: verify
21      }
22      if (data.data[0].verify == verify) {
23        if (time - data.data[0].sendtime <= data.data[0].activetime * 60 * 1000) {
24          // 验证成功
25          updataData.verifyStatus = true
26        } else {
27          code = 1001
28          msg = '验证码过期,请重新获取'
29        }
30      } else {
31        code = 1002
32        msg = '验证码错误,请重试'
33      }
34      console.log(data.data[0]._id)
35      let update = await db.collection('Verify').where({
36        '_id': data.data[0]._id
37      }).update({
38        data: updataData
39      })
40      console.log(code, msg)
41      return {
42        code,
43        msg
44      }
45    }
46  }
```

代码第13~15行查找数据库Verify集合中手机号是phone的最近一条记录。代码第22行判断用户输入的验证码和数据库查询出来的记录中的验证码是否一致。代码第23行判断当前登录的时间和发送验证码的时间差是否小于规定的时间,这里activetime为5,时间的单位为毫秒,因此代码第23行规定了登录的时间和发送验证码的时间差要小于5min,如果超过5min则会提示验证码过期,请重新获取。代码第35~39行更新数据库Verify集合中的相应的记录。

9.3 项目主页

项目主页的功能比较简单,其中一个功能是用户在线报修功能,另一个功能是管理员功能。这里为了简化,没有判断管理员的权限,实际项目中管理员功能的使用必须判断是否具有相应的权限。

项目主页 index.wxml 的代码如下：

```
1  <cu-custom bgColor="bg-gradual-pink" isBack="{{false}}">
2    <view slot="content">主页面</view>
3  </cu-custom>
4
5  <view class="cu-bar bg-white solid-bottom">
6    <view class="action">
7      <text class="cuIcon-title text-orange"></text>应用列表
8    </view>
9  </view>
10
11 <view class="cu-list grid col-3">
12   <view class="cu-item" wx:for="{{boxlist}}" wx:key="index" wx:for-item="item">
13     <view data-id='{{item.name}}' bindtap='enterapplication'>
14       <image style="width: 40px; height: 40px" src='{{item.url}}'></image>
15       <text>{{item.name}}</text>
16     </view>
17   </view>
18 </view>
```

项目主页显示效果如图 9-10 所示，包含用户在线报修功能和管理员功能。

图 9-10　报修微信小程序项目主页效果

相应的 index.js 代码如下：

```
1   const db = wx.cloud.database()
2   Page({
3     data: {
4       boxlist:
5       [
6         {
7           name:"在线报修",
8           url:"/images/repair.png"
9         },
10        {
11          name:"管理员",
12          url:"/images/admin.png"
13        }
14      ],
15    },
16
17    onLoad: function (options) {
18      var phone = wx.getStorageSync('phone');
19      if (phone == "") {
20        wx.navigateTo({
21          url: '../login/login'
22        })
23      }
24    },
25
26    enterapplication: function (event) {
27      var id = event.currentTarget.dataset.id
28      switch (id) {
29        case '在线报修':
30          wx.navigateTo({ url: '../repairtab/repairtab' })
31          break;
32        case '管理员':
33          wx.navigateTo({ url: '../repairtabadmin/repairtabadmin' })
34          break;
35        default:
36          break;
37      }
38    },
39  })
```

为了直观显示，本页面只有两个功能，通过代码第 4～15 行将这两项功能数据直接写入 data 中，实际开发中应该写入云数据库，图片也应该上传到云存储中。代码第 17～24 行的页面加载事件中，从缓存中读取手机号，如果不存在，则表示用户没有登录，跳转到登录页面。代码第 26～38 行用 switch 语句判断用户点击了哪个功能。

9.4 用户报修页面

用户报修页面 repairtab.wxml 的代码如下：

```
1   <cu-custom bgColor="bg-gradual-pink" isBack="{{true}}">
2     <view slot="backText">返回</view>
3     <view slot="content">报修单</view>
4   </cu-custom>
5
6   <view class="cu-list grid col-3">
7     <view class="cu-item">
8       <view bindtap='fillform'>
9         <image style="width:40px;height:50px" src='/images/form.png'></image>
10        <text>填写维修表</text>
11      </view>
12    </view>
13  </view>
14
15  <view class="cu-bar bg-white margin-top solid-bottom">
16    <view class="action">
17      <text class="cuIcon-title text-orange"></text>报修申请信息
18    </view>
19  </view>
20  <scroll-view scroll-x class="bg-white nav">
21    <view class="flex text-center">
22      <view class="cu-item flex-sub {{index==TabCur?'text-orange cur':''}}" wx:for="
23  {{navTab}}" wx:key bindtap="tabSelect" data-id="{{index}}">
24        {{item}}
25      </view>
26    </view>
27  </scroll-view>
28
29  <view class="cu-list menu sm-border">
30    <block wx:for="{{orderlist}}" wx:key="index">
31      <view class="cu-item">
32        <image style="width:40px;height:50px" src='/images/file.png'></image>
33        <view class="content">
34          <view class="text-df text-blue">申请人:{{item.name}}</view>
35          <view class="text-df text-blue">申请时间:{{item.time}}</view>
36          <view class="text-df text-blue">维修状态:{{item.status}}
37            <block wx:if='{{item.step==4}}'>
38              <text class="text-red">(用户评价:{{item.score}}星)</text>
39            </block>
40          </view>
41          <block wx:if="{{item.step==2}}">
42            <view class="text-df text-blue">派单时间:{{item.completetime}}</view>
43          </block>
44          <block wx:elif="{{item.step>=3}}">
45            <view class="text-df text-blue">完成时间:{{item.completetime}}</view>
46          </block>
47        </view>
48        <view class="action">
49          <view style='flex-basis:15%;justify-content:center;' bindtap="revisefile" data-index='{{index}}'>
```

```
50        data-id = '{{item._id}}'>
51            <button class = "cu-btn round bg-green ">{{item.btnText}}</button>
52          </view>
53        </view>
54      </view>
55    </block>
56 </view>
```

用户报修页面功能相对比较简单,包括用户填写报修单功能和报修申请信息(维修申请单和历史维修单)。用户报修页面效果如图9-11所示。用户报修分为4个步骤:申请报修单、报修单派单、用户评价和报修单完成。

图9-11　报修微信小程序用户报修页面效果

相应的repairtab.js代码如下:

```
1  const db = wx.cloud.database()
2  const _ = db.command
3  var util = require('../../utils/util.js')
4  Page({
5    data: {
6      TabCur: 0,
7      navTab: ['维修申请单','历史维修单'],
8      orderlist: {},
9      phone: '',
10     page: 0,
```

```javascript
11      },
12
13      LoadUnCompletedForm(isReachBottom) {
14        var page = this.data.page
15        wx.showLoading({
16          title: '数据加载中...',
17        })
18        var phone = wx.getStorageSync('phone');
19        db.collection('order').where({
20          step: _.lte(2),
21          phone: phone
22        }).skip(page).orderBy('step', 'asc')
23          .orderBy('submitdate', 'desc')
24          .get().then(res =>{
25            for (var index in res.data) {
26              res.data[index].time = util.formatTime(res.data[index].submitdate)
27              res.data[index].completetime = util.formatTime(res.data[index].completedate)
28              if (res.data[index].step == 1)
29                res.data[index].btnText = '等待派单'
30              else
31                res.data[index].btnText = '已派单'
32            }
33            var orderlist = this.data.orderlist
34            if(isReachBottom == 1)
35              orderlist = orderlist.concat(res.data)
36            else
37              orderlist = res.data
38            this.setData({
39              orderlist: orderlist
40            })
41            wx.hideLoading()
42          })
43      },
44
45      LoadCompletedForm(isReachBottom) {
46        var page = this.data.page
47        wx.showLoading({
48          title: '数据加载中...',
49        })
50        var phone = wx.getStorageSync('phone');
51        db.collection('order').where({
52          step: _.gte(3),
53          phone: phone
54        }).skip(page).orderBy('submitdate', 'desc')
55          .get().then(res =>{
56            for (var index in res.data) {
57              res.data[index].time = util.formatTime(res.data[index].submitdate)
58              res.data[index].completetime = util.formatTime(res.data[index].completedate)
59              if (res.data[index].step == 3)
```

```js
60          res.data[index].btnText = '待评价'
61        else
62          res.data[index].btnText = '已完成'
63      }
64      var orderlist = this.data.orderlist
65      if (isReachBottom == 1)
66        orderlist = orderlist.concat(res.data)
67      else
68        orderlist = res.data
69      this.setData({
70        orderlist: orderlist
71      })
72      wx.hideLoading()
73    })
74  },
75
76  onLoad: function (options) {
77    if (this.data.TabCur == 0) {
78      this.LoadUnCompletedForm(0)
79    } else if (this.data.TabCur == 1) {
80      this.LoadCompletedForm(0)
81    }
82  },
83
84  onShow: function () {
85    this.onLoad()
86  },
87
88  tabSelect(e) {
89    this.setData({
90      TabCur: e.currentTarget.dataset.id,
91      page: 0
92    })
93    if (this.data.TabCur == 0) {
94      this.LoadUnCompletedForm(0)
95    } else if (this.data.TabCur == 1) {
96      this.LoadCompletedForm(0)
97    }
98  },
99
100 onReachBottom: function () {
101   console.log("onReachBottom")
102   var page = this.data.page + 20
103   this.setData({
104     page: page
105   })
106   if (this.data.TabCur == 0) {
107     this.LoadUnCompletedForm(1)
108   } else if (this.data.TabCur == 1) {
```

```
109          this.LoadCompletedForm(1)
110        }
111      },
112  
113      fillform: function (event) {
114        wx.navigateTo({
115          url: '../orderform/orderform'
116        })
117      },
118  
119      revisefile: function (event) {
120        wx.navigateTo({
121          url: '../formdetail/formdetail?authority = 0&id = ' + event.currentTarget.dataset.id
122        })
123      },
124  })
```

维修申请信息分为维修申请单(对应 step＝1 申请报修单和 step＝2 报修单派单)和历史维修单(对应 step＝3 待评价和 step＝4 报修单完成)。代码第 13～43 行加载维修申请单(对应 step＝1 申请报修单和 step＝2 报修单派单),其中代码第 19～24 行查询数据库 order 集合中 step≤2 且 phone 等于用户手机号的记录,并按照 step 字段升序、提交时间 submitdate 降序排列;代码第 34～40 行判断是否触底加载,isReachBottom＝1 表示触底加载,需要把查询出来的时间和原有的数据进行拼接。代码第 45～74 行加载历史维修单(对应 step＝3 待评价和 step＝4 报修单完成),和前面加载维修申请单类似,查询数据库 order 集合中 step≥3 且 phone 等于用户手机号的记录。代码第 76～82 行根据选择的导航栏加载不同的记录。代码第 88～98 行对应导航栏切换事件。代码第 100～111 行对应触底加载数据。代码第 113～117 行对应用户点击页面上"填写报修单"图标,跳转到填写报修单页面 orderform。代码第 119～123 行对应单击每条报修单记录中的按钮事件,跳转到报修单页面 formdetail。

9.5 管理员管理页面

管理员管理页面和用户报修页面类似,管理员管理页面 repairtabadmin.wxml 代码如下:

```
1  <cu-custom bgColor = "bg-gradual-pink" isBack = "{{true}}">
2    <view slot = "backText">返回</view>
3    <view slot = "content">管理员管理页面</view>
4  </cu-custom>
5  
6  <view class = "cu-bar bg-white solid-bottom">
7    <view class = "action">
8      <text class = "cuIcon-title text-orange"></text>报修申请信息
9    </view>
```

```
10    </view>
11    <scroll-view scroll-x class="bg-white nav">
12      <view class="flex text-center">
13        <view class="cu-item flex-sub {{index==TabCur?'text-orange cur':''}}" wx:for="{{navTab}}" wx:key bindtap="tabSelect" data-id="{{index}}">
14    
15          {{item}}
16        </view>
17      </view>
18    </scroll-view>
19    
20    <view class="cu-list menu sm-border">
21      <block wx:for="{{orderlist}}" wx:key="index">
22        <view class="cu-item ">
23          <image style="width:40px;height:50px" src='/images/file.png'></image>
24          <view class="content">
25            <view class="text-df text-blue">申请人:{{item.name}}</view>
26            <view class="text-df text-blue">申请日期:{{item.time}}</view>
27            <view class="text-df text-blue">维修状态:{{item.status}}
28              <block wx:if='{{item.step==4}}'>
29                <text class="text-red">(用户评价:{{item.score}}星)</text>
30              </block>
31            </view>
32            <block wx:if="{{item.step==2}}">
33              <view class="text-df text-blue">派单时间:{{item.completetime}}</view>
34            </block>
35            <block wx:elif="{{item.step>=3}}">
36              <view class="text-df text-blue">完成时间:{{item.completetime}}</view>
37            </block>
38          </view>
39          <view class="action">
40            <view style='flex-basis:15%;justify-content:center;' bindtap="manageorder" data-index='{{index}}' data-id='{{item._id}}'>
41    
42              <button class="cu-btn round bg-green">{{item.btnText}}</button>
43            </view>
44          </view>
45        </view>
46      </block>
47    </view>
```

管理员管理页面的样式和用户报修页面的样式类似,这里不再重复说明。管理员管理页面的效果如图 9-12 所示。

相应的 repairtabadmin.js 代码如下:

```
1  const db = wx.cloud.database()
2  const _ = db.command
3  var util = require('../../utils/util.js')
4  Page({
5    data:{
```

图 9-12　管理员管理页面的效果

```
6      TabCur: 0,
7      navTab: ['维修申请单', '历史维修单'],
8      orderlist: {},
9      btnText: '',
10     page: 0
11   },
12
13   LoadUnCompletedForm(isReachBottom) {
14     var page = this.data.page
15     wx.showLoading({
16       title: '数据加载中...',
17     })
18     db.collection('order').where({
19       step: _.lte(2),
20     }).skip(page).orderBy('step', 'asc')
21       .orderBy('submitdate', 'desc')
22       .get().then(res =>{
23         for (var index in res.data) {
24           res.data[index].time = util.formatTime(res.data[index].submitdate)
25           res.data[index].completetime = util.formatTime(res.data[index].completedate)
26           if (res.data[index].step == 1)
27             res.data[index].btnText = '等待派单'
28           else
29             res.data[index].btnText = '已派单'
30         }
```

```js
      var orderlist = this.data.orderlist
      if (isReachBottom == 1)
        orderlist = orderlist.concat(res.data)
      else
        orderlist = res.data
      this.setData({
        orderlist: orderlist
      })
      wx.hideLoading()
    })
  },

  LoadCompletedForm(isReachBottom) {
    var page = this.data.page
    wx.showLoading({
      title: '数据加载中...',
    })
    db.collection('order').where({
      step: _.gte(3),
    }).skip(page).orderBy('submitdate', 'desc')
      .get().then(res =>{
        for (var index in res.data) {
          res.data[index].time = util.formatTime(res.data[index].submitdate)
          res.data[index].completetime = util.formatTime(res.data[index].completedate)
          if (res.data[index].step == 3)
            res.data[index].btnText = '待评价'
          else
            res.data[index].btnText = '已完成'
        }
        var orderlist = this.data.orderlist
        if (isReachBottom == 1)
          orderlist = orderlist.concat(res.data)
        else
          orderlist = res.data
        this.setData({
          orderlist: orderlist
        })
        wx.hideLoading()
      })
  },

  onLoad: function (options) {
    if (this.data.TabCur == 0) {
      this.LoadUnCompletedForm(0)
    } else if (this.data.TabCur == 1) {
      this.LoadCompletedForm(0)
    }
  },
```

```
80      onShow: function () {
81        this.onLoad()
82      },
83
84      tabSelect(e) {
85        this.setData({
86          TabCur: e.currentTarget.dataset.id, page: 0
87        })
88        if (this.data.TabCur == 0) {
89          this.LoadUnCompletedForm(0)
90        } else if (this.data.TabCur == 1) {
91          this.LoadCompletedForm(0)
92        }
93      },
94
95      onReachBottom: function () {
96        console.log("onReachBottom")
97        var page = this.data.page + 20
98        this.setData({
99          page: page
100       })
101       if (this.data.TabCur == 0) {
102         this.LoadUnCompletedForm(1)
103       } else if (this.data.TabCur == 1) {
104         this.LoadCompletedForm(1)
105       }
106     },
107
108     managerorder: function (event) {
109       wx.navigateTo({
110         url: '../formdetail/formdetail?authority=1&id=' + event.currentTarget.dataset.id
111       })
112     }
113   })
```

同样 repairtabadmin.js 代码和 repairtab 代码类似，区别在于 repairtab 查询数据库 order 集合中手机字段为用户手机号的报修单记录，repairtabadmin 查询所有用户的报修单记录。

9.6 填写报修单页面

填写报修单页面（pages/orderform/orderform）样式采用 ColorUI 中的表单样式（代码见 pages/component/form），orderform.wxml 代码如下：

```
1   <cu-custom bgColor="bg-gradual-pink" isBack="{{true}}">
2     <view slot="backText">返回</view>
3     <view slot="content">报修申请表</view>
4   </cu-custom>
5
```

```
6    <form>
7      <view class="cu-form-group">
8        <view class="title">
9          <text style="color:red">*</text>报修人:</view>
10         <input bindinput="inputname" placeholder="请输入姓名" value='{{name}}'></input>
11     </view>
12     <view class="cu-form-group">
13       <view class="title">
14         <text style="color:red">*</text>手机号码:</view>
15         <input placeholder="手机号码" value='{{phone}}'></input>
16     </view>
17     <view class="cu-form-group">
18       <view class="title">
19         <text style="color:red">*</text>详细地点:</view>
20         <input bindinput="inputplace" placeholder="例如*栋***房间" value='{{place}}'></input>
21     </view>
22     <view class="cu-form-group align-start">
23       <view class="title">
24         <text style="color:red">*</text>故障现象:</view>
25         <textarea maxlength="-1" bindinput="inputdemand" placeholder="输入故障现象" value=
26  '{{symptom}}'></textarea>
27     </view>
28
29     <view class="cu-bar bg-white margin-top">
30       <view class="action">
31         图片上传
32       </view>
33       <view class="action">
34         {{imgList.length}}/3
35       </view>
36     </view>
37     <view class="cu-form-group">
38       <view class="grid col-3 grid-square flex-sub">
39         <view class="bg-img" wx:for="{{imgList}}" wx:key="{{index}}" bindtap="ViewImage"
40  data-url="{{imgList[index]}}">
41           <image src='{{imgList[index]}}' mode='aspectFill'></image>
42           <view class="cu-tag bg-red" catchtap="DelImg" data-id="{{index}}">
43             <text class="cuIcon-close"></text>
44           </view>
45         </view>
46         <view class="solids" bindtap="ChooseImage" wx:if="{{imgList.length<3}}">
47           <text class="cuIcon-cameraadd"></text>
48         </view>
49       </view>
50     </view>
51   </form>
52
53   <view class="flex padding justify-center">
54     <button class="cu-btn lg bg-pink" bindtap="submitform">提交</button>
55   </view>
```

代码第 7~27 行对应表单中的输入框。代码第 29~50 行对应图片上传控件（采用 ColorUI 表单中的样式），在本项目中允许用户最多上传 3 张图片。填写报修单页面的效果如图 9-13 所示，用户可以在上传的图片右上角点击"×"按钮，删除上传的图片，也可以点击图片进行全屏浏览。

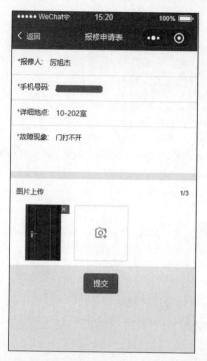

图 9-13　报修微信小程序填写报修单页面的效果

相应的 order.js 代码如下：

```
1   var util = require('../../utils/util.js')
2   Page({
3     data: {
4       name: '',
5       place: '',
6       phone: '',
7       symptom: '',
8       imgList: []
9     },
10
11    onLoad: function (options) {
12      var phone = wx.getStorageSync('phone');
13      this.setData({ phone: phone })
14    },
15
16    inputname: function (event) {
17      this.setData({
18        name: event.detail.value
```

```
19        })
20     },
21
22     inputplace: function (event) {
23        this.setData({
24           place: event.detail.value
25        })
26     },
27
28     inputdemand: function (event) {
29        this.setData({
30           symptom: event.detail.value
31        })
32     },
33
34     ChooseImage() {
35        wx.chooseImage({
36           count: 3,                                    //默认为9
37           sizeType: ['original', 'compressed'],        //可以指定是原图还是压缩图,默认二者都有
38           sourceType: ['album', 'camera'],
39           success: (res) =>{
40              wx.showLoading({
41                 title: '图片上传中...',
42              })
43              var time = util.formatTimestring(new Date());
44              var FilePaths = []
45              const temFilePaths = res.tempFilePaths
46              let promiseArr = [];
47              for (let i = 0; i < temFilePaths.length; i++) {
48                 let promise = new Promise((resolve, reject) =>{
49                    var randstring = Math.floor(Math.random() * 1000000).toString() + '.png'
50                    randstring = time + '-' + randstring
51                    wx.cloud.uploadFile({
52                       cloudPath: 'orderimg/' + randstring,
53                       filePath: temFilePaths[i],        //文件路径
54                       success: res =>{
55                          // get resource ID
56                          console.log(res.fileID)
57                          FilePaths[i] = res.fileID
58                          resolve(res);
59                       },
60                       fail: err =>{
61                          reject(error);
62                       }
63                    })
64                 })
65                 promiseArr.push(promise)
66              }
67              Promise.all(promiseArr).then((result) =>{
```

```
68              if(this.data.imgList.length != 0) {
69                this.setData({
70                  imgList: this.data.imgList.concat(FilePaths)
71                })
72              } else {
73                this.setData({
74                  imgList: FilePaths
75                })
76              }
77              wx.hideLoading()
78            })
79          }
80        });
81      },
82
83      ViewImage(e) {
84        wx.previewImage({
85          urls: this.data.imgList,
86          current: e.currentTarget.dataset.url
87        });
88      },
89
90      DelImg: function (event) {
91        console.log(event.currentTarget.dataset.id)
92        var id = event.currentTarget.dataset.id
93        var imgList = this.data.imgList
94        wx.cloud.deleteFile({
95          fileList: [imgList[id]]
96        }).then(res =>{
97          imgList.splice(id, 1)
98          this.setData({ imgList: imgList })
99          console.log(res.fileList)
100       }).catch(error =>{
101       })
102     },
103
104     check() {
105       var name = this.data.name
106       var place = this.data.place
107       var symptom = this.data.symptom
108       if (name == '' || place == '' || symptom == '') {
109         return false
110       }
111       else {
112         return true
113       }
114     },
115
116     submitform: function (event) {
```

```
117       var userInfo = wx.getStorageSync('userInfo');
118       if (this.check()) {
119         wx.cloud.callFunction({
120           name: 'orderform',
121           data: {
122             $url: "add",
123             name: this.data.name,
124             phone: this.data.phone,
125             place: this.data.place,
126             symptom: this.data.symptom,
127             url: this.data.imgList,
128             avatarUrl: userInfo.avatarUrl
129           },
130           success: res =>{
131             wx.showToast({
132               title: '报修单提交成功',
133               icon: 'success',
134               duration: 2000
135             })
136             wx.navigateBack({
137               delta: 1
138             })
139           }
140         })
141       }
142       else {
143         wx.showToast({
144           title: '信息填写不全!',
145           icon: 'success',
146           duration: 2000
147         })
148       }
149     },
150   })
```

代码第 3~9 行对应 data 数据项：name 变量为报修人姓名；place 变量为详细地点；phone 变量为报修人手机号，在这里不用用户输入，提取用户登录页的手机号码；symtom 变量为故障现象；imgList 对应上传的图片数组。代码第 11~14 行对应页面加载事件中读取缓存中用户的手机号码。代码第 16~32 行对应用户输入报修人、详细地点和故障现象。代码第 34~81 行对应用户上传图片，ChooseImage()从相册中选取少于 3 张图片，为了防止图片冲突，上传的图片以日期＋6 位随机数作为文件名，通过 wx.cloud.uploadFile 上传图片，由于 wx.cloud.uploadFile 操作是异步的，在这种情况下图片的上传完成后输出的顺序、判断图片是否全部上传完成等，都是一些需要注意的地方。Promise.all()将多个 Promise 实例包装成一个新的 Promise 实例(简单来说就是当所有的异步请求成功后才会执行)，在这个方法的帮助下，这些问题都迎刃而解。代码第 67~78 行通过 Promise.all() 的方法，把用户上传的多张图片放到 imgList 中。代码第 83~88 行对应全屏查看用户上传的图片。代码第 90～102 行对应用户点击图片右上角的"×"按钮，使用 wx.cloud.deleteFile 对云存储上的照片进行删除。代码第 104～114 行检查用户是否填写用户名、详细地点以及故障现象。代码第 116～149 行把用户填写的信息以及上传图片的地址写入云数据库 order 集合中。代码第 119～129 行把这些数据通过云函数 orderform 添加到数据库 order 集合中，提交报修单后 order 集合记录如图 9-14 所示。

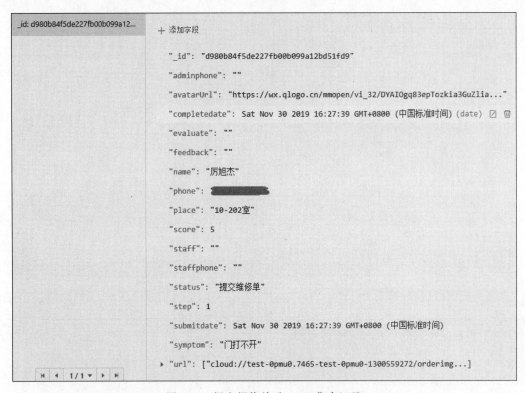

图 9-14　提交报修单后 order 集合记录

云函数 orderform 的代码如下：

```
1   const cloud = require('wx-server-sdk')
2   cloud.init()
3   const TcbRouter = require('tcb-router');
4   const db = cloud.database()
5   const _ = db.command
6   exports.main = async (event, context) =>{
7     const app = new TcbRouter({ event });
8     //路由为字符串,该中间件只适用于 add 路由
9     app.router('add', async (ctx, next) =>{
10      try {
11        return await db.collection('order').add({
12          data: {
13            name: event.name,
14            phone: event.phone,
15            place: event.place,
```

```js
          symptom: event.symptom,
          url: event.url,
          avatarUrl: event.avatarUrl,
          submitdate: new Date(),
          completedate: new Date(),
          status: '提交维修单',
          step: 1,
          staff: '',
          staffphone: '',
          adminphone: '',
          feedback: '',
          score: 5,
          evaluate: ''
        }
    })
  } catch (e) {
    console.error(e)
  }
  await next(); //执行下一中间件
})

//路由为字符串,该中间件只适用于 update_step2 路由
app.router('update_step2', async (ctx, next) =>{
  try {
    return await db.collection('order').doc(event.id).update({
      data: {
        staff: event.staff,
        staffphone: event.staffphone,
        adminphone: event.adminphone,
        feedback: event.feedback,
        step: 2,
        completedate: new Date(),
        status: "已派单"
      }
    })
  } catch (e) {
    console.error(e)
  }
  await next(); // 执行下一中间件
})

//路由为字符串,该中间件只适用于 update_step3 路由
app.router('update_step3', async (ctx, next) =>{
  try {
    return await db.collection('order').doc(event.id).update({
      data: {
        adminphone: event.adminphone,
        step: 3,
        completedate: new Date(),
```

```
65              status:"待评价"
66          }
67        })
68      } catch (e) {
69        console.error(e)
70      }
71      await next();          //执行下一中间件
72    })
73
74    //路由为字符串,该中间件只适用于 update_step4 路由
75    app.router('update_step4', async (ctx, next) =>{
76      try {
77        return await db.collection('order').doc(event.id).update({
78          data: {
79            step: 4,
80            status:"已完成",
81            score: event.score,
82            evaluate: event.evaluate
83          }
84        })
85      } catch (e) {
86        console.error(e)
87      }
88      await next();          //执行下一中间件
89    })
90
91    return app.serve();
92  }
```

每个报修单分为 4 个步骤：step=1 申请报修单、step=2 报修单派单、step=3 待评价和 step=4 报修单完成。代码第 8～35 行对应提交申请报修单,向数据库 order 集合中添加记录。代码第 37～55 行对应报修单派单,数据库 order 集合中相应报修单更新 step=2,并添加报修单派修信息。代码第 57～72 行对应修改报修单为待评价状态 step=3。代码第 74～89 行对应修改报修单为报修单完成状态 step=4。填写完报修单后进入用户报修页面可以看到刚提交的维修单,如图 9-15 所示。

9.7 报修单页面

报修微信小程序分为 4 个步骤：步骤 1,用户填写报修单；步骤 2,管理员对报修单进行派

图 9-15 提交报修记录后用户报修页面效果

单;步骤3,维修完成后,管理员修改报修单状态为报修单完成;步骤4,用户对报修单进行用户评价。9.6节中填写报修单页面已经完成了报修单填写,也就是向数据库order集合中添加了一条报修记录,接下来的步骤2~4将在报修单页面完成。

报修单中使用了ColorUI中的步骤条(代码见pages/component/step),如图9-16所示。

报修单中使用了Vant Weapp中的Rate评分(https://youzan.github.io/vant-weapp/#/rate),如图9-17所示。

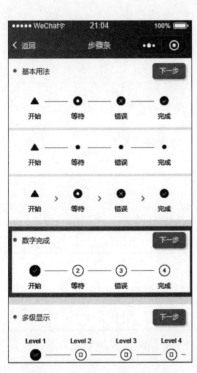

图9-16 ColorUI中的步骤条

图9-17 Vant Weapp中的Rate评分效果

报修单formdetail.wxml代码如下:

```
1   <cu-custom bgColor="bg-gradual-pink" isBack="{{true}}">
2       <view slot="backText">返回</view>
3       <view slot="content">维修单详情</view>
4   </cu-custom>
5
6   <block wx:if="{{order.url.length>0}}">
7       <swiper class="screen-swiper {{DotStyle?'square-dot':'round-dot'}}" indicator-dots="true"
8   circular="true" autoplay="true" interval="5000" duration="500">
9           <swiper-item wx:for="{{order.url}}" wx:key="img" bindtap="showImg" id="img"
```

```
10        data-imgs='{{order.url}}' data-currentimg="{{img}}">
11           <image src="{{item}}" mode="scaleToFill"></image>
12        </swiper-item>
13     </swiper>
14  </block>
15
16  <view class="cu-card dynamic no-card">
17     <view class="cu-bar bg-white solid-bottom">
18        <view class="action">
19           <text class="cuIcon-title text-orange"></text>维修单状态({{order.status}})
20        </view>
21     </view>
22     <view class="bg-white padding">
23        <view class="cu-steps">
24           <view class="cu-item {{index>order.step-1?'':'text-blue'}}" wx:for="{{numList}}" wx:key>
25              <text class="num" data-index="{{index+1}}"></text>{{item.name}}
26           </view>
27        </view>
28     </view>
29
30     <view class="cu-bar bg-white solid-bottom margin-top">
31        <view class="action">
32           <text class="cuIcon-title text-orange"></text>维修单进度
33        </view>
34     </view>
35     <view class="cu-list menu-avatar">
36        <view class="cu-item">
37           <view class="cu-avatar round lg" style="background-image:url({{order.avatarUrl}});"></view>
38           <view class="content flex-sub">
39              <view>{{order.name}}</view>
40              <view class="text-gray text-sm flex justify-between">
41                 {{order.time}}
42              </view>
43           </view>
44        </view>
45     </view>
46     <view class="cu-form-group">
47        <view class="title">
48           <text style="color:red">*</text>手机号码:</view>
49        <input placeholder="输入手机号码" value='{{order.phone}}' disabled></input>
50     </view>
51     <view class="cu-form-group">
52        <view class="title">
53           <text style="color:red">*</text>详细地点:</view>
54        <input placeholder="输入详细地点" value='{{order.place}}' disabled></input>
55     </view>
56     <view class="cu-form-group align-start">
57        <view class="title">
58           <text style="color:red">*</text>故障现象:</view>
```

```
59      <textarea maxlength="-1" placeholder="输入故障现象" value='{{order.symptom}}'></textarea>
60    </view>
61
62    <view class="cu-bar bg-white solid-bottom margin-top">
63      <view class="action">
64        <text class="cuIcon-title text-orange"></text>维修信息-管理员填写
65      </view>
66    </view>
67    <view class="cu-form-group">
68      <view class="title">
69        <text style="color:red">*</text>维修人员:</view>
70      <input bindinput="inputstaff" placeholder="输入维修人员" value='{{order.staff}}'
71 disabled='{{authority==0}}'></input>
72    </view>
73    <view class="cu-form-group">
74      <view class="title">
75        <text style="color:red">*</text>维修人员电话:</view>
76      <input bindinput="inputstaffphone" placeholder="维修人员电话" value='{{order.staffphone}}'
77 disabled='{{authority==0}}'></input>
78    </view>
79    <view class="cu-form-group align-start">
80      <view class="title">
81        维修反馈:</view>
82      <textarea maxlength="-1" bindinput="inputfeedback" placeholder="输入反馈"
83 value='{{order.feedback}}'></textarea>
84    </view>
85
86    <block wx:if="{{order.step<=2&&authority==1}}">
87      <view class="flex padding justify-center">
88        <button class="cu-btn lg bg-pink" bindtap="assignform">派单</button>
89        <button class="cu-btn lg bg-pink" bindtap="completeform">完成</button>
90      </view>
91    </block>
92
93    <block wx:if="{{order.step>=3}}">
94      <view class="cu-bar bg-white solid-bottom margin-top">
95        <view class="action">
96          <text class="cuIcon-title text-orange"></text>用户评价
97        </view>
98      </view>
99      <view class="cu-form-group">
100       <view class="title">
101         用户评分:</view>
102       <van-rate value="{{order.score}}" count="{{5}}" disabled type="" color="#f44"
103 bind:change="onChange" disabled='{{authority==1||order.step==4}}' />
104     </view>
105     <view class="cu-form-group align-start">
106       <view class="title">
107         用户评价:</view>
```

```
108        <textarea maxlength="-1" bindinput="inputevaluate" placeholder="输入反馈" value=
109    '{{order.evaluate}}'></textarea>
110      </view>
111
112      <block wx:if="{{authority==0&&order.step==3}}">
113        <view class="flex padding justify-center">
114          <button class="cu-btn lg bg-pink" bindtap="evaluateform">提交</button>
115        </view>
116      </block>
117    </block>
118  </view>
```

代码第 6～14 行显示报修单图片轮播图。代码第 16～28 行采用了 ColorUI 中的步骤条，总共分为 4 个步骤：申请报修单、报修单派单、报修单完成和用户评价。代码第 30～60 行对应报修单进度，也就是用户填写报修单的数据，只是现在输入框的状态变成了 disable 禁用状态。代码第 62～84 行对应管理员派单输入信息，包括维修人员姓名、维修人员电话、维修反馈。代码第 86～91 行对应单击"派单"按钮和"完成派修单"按钮，不过这 2 个按钮只会出现在步骤 1 和步骤 2 且对应管理员权限 authority==1 时才会显示，普通用户是无法看到这 2 个按钮的。代码第 94～104 行对应用户评价部分，用户评分只能在步骤 3 且是普通用户才能评分，其他情况下评分状态是禁用的。因为报修单页面包含了步骤 2、步骤 3 和步骤 4，而且要区分普通用户和管理员不同的功能，所以页面中有比较多的变量控制部分功能的显示与隐藏、开启与禁用，读者需要根据需求仔细研究。在报修单处于 step＝1 的情况下，普通用户进入报修单页面的显示效果如图 9-18 所示。

在报修单处于 step＝1 的情况下，管理员进入报修单页面的显示效果如图 9-19 所示。管理员可以输入维修信息，可以在该页面进行派单和完成报修单。

图 9-18 普通用户报修单 step＝1 效果图

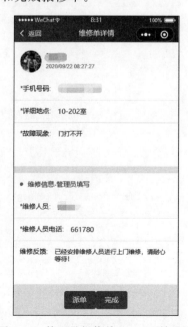
图 9-19 管理员报修单 step＝1 效果图

管理员单击"派单"按钮后,维修单步骤变成 step=2,这时用户就会收到一条报修单派修短信,如图 9-20 所示。

接下来就是等维修工人进行上门服务,等维修完成后,管理员进入报修单单击"完成"按钮,维修单步骤变成 step=3,同样用户会收到一条报修单完成的短信,如图 9-21 所示。

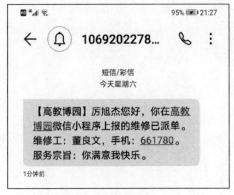 

图 9-20　报修单派修短信　　　　　　图 9-21　报修单完成短信

最后一个步骤就是用户进行评价,用户进入用户报修页面,选择"历史维修单"选项,对报修单进行评价,如图 9-22 所示。

图 9-22　用户进行评价

相应的 formdetail.js 代码如下:

```
var util = require('../../utils/util.js'
const db = wx.cloud.database()
```

```js
Page({
  data: {
    order: {},
    orderid: '',
    authority: 0,
    current: 0,
    numList: [{
      name: '申请报修单'
    }, {
      name: '报修单派单'
    }, {
      name: '报修单完成'
    }, {
      name: '用户评价'
    }, ],
  },

  onLoad: function(options) {
    console.log(options)
    const orderid = options.id
    const authority = options.authority
    db.collection('order').doc(orderid).get().then(res =>{
      res.data.time = util.formatTime(res.data.submitdate)
      this.setData({
        orderid: orderid,
        authority: authority,
        order: res.data
      })
    })
  },

  showImg: function(event) {
    var that = this;
    console.log(event);
    var imgs = event.currentTarget.dataset.imgs;
    var temp = [];
    for (var index in imgs) {
      temp = temp.concat(imgs[index]);
    }
    wx.previewImage({
      current: event.currentTarget.dataset.currentimg,
      urls: temp,
    })
  },

  onChange(event) {
    var order = this.data.order
    order.score = event.detail
    this.setData({
```

```js
52        order: order
53      });
54    },
55
56    inputstaff: function(event) {
57      var order = this.data.order
58      order.staff = event.detail.value
59      this.setData({
60        order: order
61      })
62    },
63
64    inputstaffphone: function(event) {
65      var order = this.data.order
66      order.staffphone = event.detail.value
67      this.setData({
68        order: order
69      })
70    },
71
72    inputfeedback: function(event) {
73      var order = this.data.order
74      order.feedback = event.detail.value
75      this.setData({
76        order: order
77      })
78    },
79
80    inputevaluate: function(event) {
81      var order = this.data.order
82      order.evaluate = event.detail.value
83      this.setData({
84        order: order
85      })
86    },
87
88    check() {
89      var staff = this.data.order.staff
90      var staffphone = this.data.order.staffphone
91      if (staff == '' || staffphone == '') {
92        return false
93      } else {
94        return true
95      }
96    },
97
98    assignform: function(event) {
99      if (this.check()) {
100       wx.showLoading({
```

```js
        title: '数据加载中...',
      })
      var phone = wx.getStorageSync('phone');
      wx.cloud.callFunction({
        name: 'orderform',
        data: {
          $url: "update_step2",
          mode: 'updateassign',
          id: this.data.order._id,
          staff: this.data.order.staff,
          staffphone: this.data.order.staffphone,
          feedback: this.data.order.feedback,
          adminphone: phone
        },
        success: res =>{
          wx.cloud.callFunction({
            name: 'sendsms',
            data: {
              phone: this.data.order.phone,
              templateId: 386955,
              name: this.data.order.name,
              staff: this.data.order.staff,
              staffphone: this.data.order.staffphone,
            },
            success: res =>{
              wx.hideLoading()
              wx.showToast({
                title: '派单成功',
                icon: 'success',
                duration: 2000
              })
              wx.navigateBack({
                delta: 1
              })
            }
          })
        }
      })
    } else {
      wx.showToast({
        title: '信息填写不全!',
        icon: 'success',
        duration: 2000
      })
    }
  },

  completeform: function(event) {
    var phone = wx.getStorageSync('phone');
```

```
150        wx.showLoading({
151          title: '数据加载中...',
152        })
153        wx.cloud.callFunction({
154          name: 'orderform',
155          data: {
156            $url: "update_step3",
157            id: this.data.order._id,
158            adminphone: phone
159          },
160          success: res =>{
161            wx.cloud.callFunction({
162              name: 'sendsms',
163              data: {
164                phone: this.data.order.phone,
165                name: this.data.order.name,
166                templateId: 386954,
167              },
168              success: res =>{
169                wx.hideLoading()
170                wx.showToast({
171                  title: '报修单完成',
172                  icon: 'success',
173                  duration: 2000
174                })
175                wx.navigateBack({
176                  delta: 1
177                })
178              }
179            })
180          }
181        })
182      },
183
184      evaluateform: function(event) {
185        wx.cloud.callFunction({
186          name: 'orderform',
187          data: {
188            $url: "update_step4",
189            id: this.data.order._id,
190            score: this.data.order.score,
191            evaluate: this.data.order.evaluate
192          },
193          success: res =>{
194            wx.showToast({
195              title: '评价完成',
196              icon: 'success',
197              duration: 2000
198            })
```

```
199          wx.navigateBack({
200            delta: 1
201          })
202        }
203      })
204    },
205  })
```

代码第 20~32 行对应页面加载事件，读取报修单记录，并获取用户权限。代码第 34~46 行实现用户单击轮播图中的图片后，实现全屏浏览。代码第 48~54 行对应用户评分。代码第 56~62 行对应管理员输入维修人员姓名。代码第 64~70 行对应管理员输入维修人员电话号码。代码第 72~78 行对应管理员输入维修反馈。代码第 80~86 行对应用户输入用户评价。代码第 88~96 行检测管理员是否输入维修人员姓名和维修人员电话号码。代码第 98~146 行对应单击"管理员派单"按钮，其中代码第 104~114 行更新报修单步骤 step＝2；代码第 115~124 行执行派单成功后给用户发送派单成功短信。代码第 148~182 行对应管理员单击"完成订单"按钮，其中第 153~159 行更新报修单步骤 step＝3；代码第 161~167 行执行维修单完成后给用户发送维修完成短信。代码第 184~204 行对应对维修单进行单击"用户评价"按钮，更新报修单步骤 step＝4。

用户评价维修单后，最终数据库 order 集合中报修单的数据如图 9-23 所示。

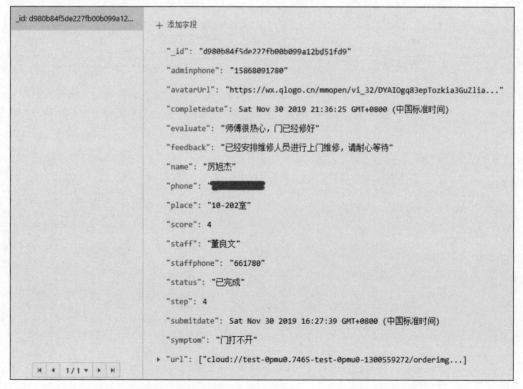

图 9-23　用户评价维修单后最终数据库 order 集合中报修单的数据

第10章

网上书城微信小程序

网上书城微信小程序包含 8 个页面：

```
1   "pages": [
2     "pages/home/home",              //项目主页
3     "pages/login/login",            //登录页面
4     "pages/addbook/addbook",        //添加图书页面
5     "pages/books/books",            //图书信息页面
6     "pages/bookdetail/bookdetail",  //图书详情页面
7     "pages/shoppingcart/shoppingcart", //购物车页面
8     "pages/order/order",            //提交订单页面
9     "pages/myorder/myorder"         //我的订单页面
10  ],
```

其中，包含 4 个数据库集合：Verify(存储短信验证码等登录信息)、book(存储商城图书信息)、shoppingcast(购物车)和 order(存储提交的订单)。权限均为"所有用户可读，仅创建者可读写"。运行本项目程序需要先在数据库中创建这 4 个集合。

本项目案例使用的登录页面和第 9 章报修类微信小程序使用的登录页面一样，在此不再重复说明。具体操作详见下方二维码。

第 11 章

团购类微信小程序

团购类微信小程序包含 7 个页面：

```
1  "pages": [
2      "pages/home/home",                    //项目主页
3      "pages/auth/auth",                    //登录页面
4      "pages/classification/classification", //商品分类页面
5      "pages/gooddetail/gooddetail",        //商品详情页面
6      "pages/shoppingcart/shoppingcart",    //购物车页面
7      "pages/order/order",                  //提交订单页面
8      "pages/myorder/myorder"               //我的订单页面
9  ],
```

其中，包含 4 个数据库集合 classification(存储商品类别)、goods(存储商品信息)、order(存储提交的订单)和 shoppingcast(购物车)。权限均为"所有用户可读，仅创建者可读写"。要运行本项目程序需要先在数据库中创建这 4 个集合。为了方便运行演示项目，本书提供了 classification 和 goods 的数据以及图片资源，用户可以把 classification.json 和 goods.json 文件直接导入数据库集合中，导入后的 classification 集合和 goods 集合如图 11-1 和图 11-2 所示。

图 11-1　数据库 classification 集合

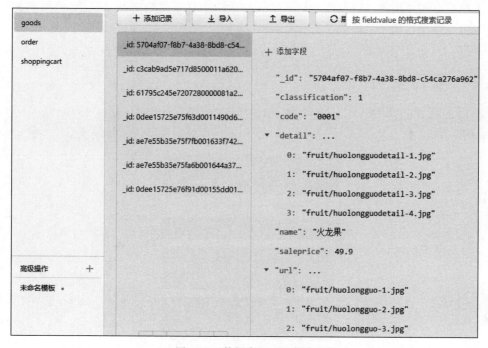

图 11-2 数据库 goods 集合

项目提供了部分商品的图片，用户需要把图片上传到自己的云存储中，进入云开发控制台，选择"存储管理"选项卡，新建 image 文件夹，云存储中图片资源如图 11-3 所示。在 image 文件夹中新建 classification 文件夹存储商品类别的图片，drink 文件夹存储茶水饮料商品的图片，fruit 文件夹存储新鲜水果商品的图片，nut 文件夹存储坚果炒货商品的图片，office 文件夹存储办公用品商品的图片，然后上传本项目提供的图片到相应的文件夹下。

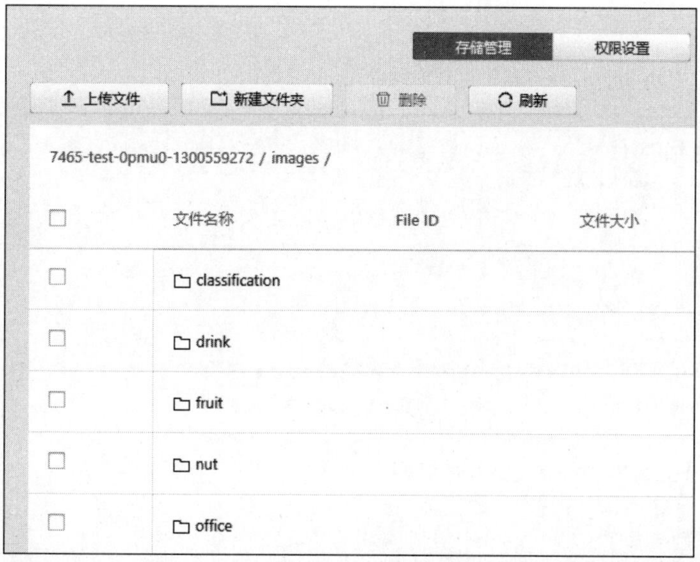

图 11-3 云存储中图片资源

本项目案例使用的登录页面和第 7 章投票微信小程序使用的登录页面一样,在此不再重复说明。登录界面效果如图 7-1 所示。

11.1 项目主页

项目主页样式采用 ColorUI 中的 Grid 布局样式(代码见 pages/basics/layout),如图 11-4 所示。

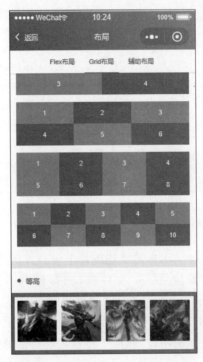

图 11-4　ColorUI 中的 Grid 布局样式

本案例项目主页 home.wxml 的代码如下:

```
1   <cu-custom bgColor = "bg-gradual-pink" isBack = "{{false}}">
2     <view slot = "backText">返回</view>
3     <view slot = "content">首页</view>
4   </cu-custom>
5
6   <view class = "bg-white padding">
7     <view class = "grid col-4">
8       <view wx:for = "{{classification}}" wx:key = "index">
9         <view bindtap = "selectclass" data-index = "{{index}}">
10          <image style = "width:180rpx;height:180rpx; border-radius:50%"
11  src = "{{'cloud://test-0pmu0.7465-test-0pmu0-1300559272/images/' + item.url }}"
12  mode = "aspectFill"></image>
13        </view>
14        <view style = "display: flex;flex-direction: column; align-items: center;">{{item.name}}</view>
```

```
15        </view>
16      </view>
17    </view>
18
19    <block wx:for="{{goodinfo}}" wx:for-item="gooditem" wx:key="index">
20      <view class="bg-white margin-top">
21        <view class="grid col-3">
22          <view class="bg-img" wx:for="{{gooditem.url.length<3?gooditem.url.length:3}}" wx:key="index">
24            <image style="width:220rpx;height:200rpx;"
25  src="{{'cloud://test-0pmu0.7465-test-0pmu0-1300559272/images/'+gooditem.url[index]}}"
26  mode="aspectFill"></image>
27          </view>
28        </view>
29        <view style="display: flex;">
30          <view>
31            <view>{{gooditem.name}}</view>
32            <view class="text-red text-xxl">
33              ¥{{gooditem.saleprice}}元
34            </view>
35          </view>
36          <view class="flex-treble flex align-center justify-end padding-sm">
37            <button class="cu-btn round bg-red shadow" bindtap="addtoCart"
38  data-code="{{gooditem.code}}" data-index="{{index}}">加入购物车</button>
39          </view>
40        </view>
41      </view>
42    </block>
```

项目主页效果如图11-5所示。页面分为两部分：上半部分显示商品分类，本项目把商品分为8类，用户可以单击商品分类的图片；下半部分显示相应的商品记录。代码第6～17行对应商品分类布局，本项目采用ColorUI中的Grid布局，每行显示4张图片，需要说明的是数据库中存入的图片地址是在云存储中的图片文件夹的位置，读者需要把第10～12行中image组件中的src地址换为自己开发环境图片的Field ID(文件ID)。第19～42行同样采用ColorUI中的Grid布局，每行显示3张图片。

相应的home.js代码如下：

```
1   const db = wx.cloud.database()
2   var app = getApp()
3   const _ = db.command
4   Page({
5     data: {
6       classification: [],
7       goodinfo: [],
8       class_cur: 1
9     },
10
```

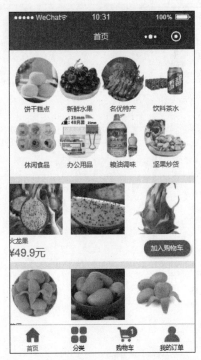

图 11-5　团购类微信小程序项目主页效果

```
11    onLoad: function(options) {
12      db.collection('classification').orderBy('id', 'asc').get().then(res =>{
13        this.setData({
14          classification: res.data
15        })
16      })
17      db.collection('goods').where({
18        classification: this.data.class_cur
19      }).get().then(res =>{
20        this.setData({
21          goodinfo: res.data
22        })
23      })
24      var userInfo = wx.getStorageSync('userInfo');
25      if (userInfo != "") {
26        db.collection('shoppingcart').where({
27          _openid: userInfo._openid
28        }).get().then(res =>{
29          var record = res.data
30          var sum = 0
31          for (let i = 0; i < record.length; i++) {
32            sum = sum + record[i].num
33          }
34          app.globalData.goodnum = sum
35          if (sum != 0) {
```

```
36            wx.setTabBarBadge({
37              index: 2,
38              text: sum + ""
39            })
40          } else {
41            wx.removeTabBarBadge({
42              index: 2
43            })
44          }
45        })
46      }
47    },
48
49    selectclass: function(event) {
50      var class_cur = this.data.class_cur
51      var index = event.currentTarget.dataset.index
52      if (class_cur != index) {
53        db.collection('goods').where({
54          classification: index
55        }).get().then(res =>{
56          this.setData({
57            goodinfo: res.data,
58            class_cur: index
59          })
60        })
61      }
62    },
63
64    addtoCart: function(event) {
65      var userInfo = wx.getStorageSync('userInfo');
66      if (userInfo == "") {
67        wx.navigateTo({
68          url: '../auth/auth'
69        })
70      } else {
71        var index = event.currentTarget.dataset.index
72        var code = event.currentTarget.dataset.code
73        var item = this.data.goodinfo[index]
74        item.submitdate = db.serverDate()
75        item.openid = userInfo.openid
76        item.num = 1
77        delete item._id
78        db.collection('shoppingcart').where({
79          _openid: userInfo._openid,
80          code: code
81        }).get().then(res =>{
82          var cart = res.data
83          if (cart.length == 0) {
84            db.collection('shoppingcart').add({
```

```
 85              data: item,
 86            })
 87            .then(res =>{
 88              var sum = app.globalData.goodnum + 1
 89              app.globalData.goodnum = sum
 90              wx.setTabBarBadge({
 91                index: 2,
 92                text: sum + ""
 93              })
 94            })
 95          } else {
 96            db.collection('shoppingcart').doc(cart[0]._id).update({
 97              data: {
 98                num: _.inc(1)
 99              }
100            }).then(res =>{
101              var sum = app.globalData.goodnum + 1
102              app.globalData.goodnum = sum
103              wx.setTabBarBadge({
104                index: 2,
105                text: sum + ""
106              })
107            })
108          }
109        })
110      }
111    }
112  })
```

代码第 11～47 行对应页面加载事件。第 12～16 行获取 classification 数据库集合中的数据。第 17～23 行获取 goods 数据库集合中的数据。第 24～46 行统计购物车中商品的数量并显示在 tabBar 中购物车的右上角,这里读取缓存中 userInfo 的数据,如果用户已经登录,则根据用户的_openid 字段查询 shoppingcart 数据库集合中的记录,把购物车中该用户的商品数量存放到全局变量 app.globalData.goodnum 中,便于其他页面调用。wx.setTabBarBadge 为 tabBar 某一项的右上角添加文本,wx.removeTabBarBadge 对应移除 tabBar 某一项右上角的文本。第 49～62 行对应用户单击分类图片,从 goods 数据库集合中查询对应分类的商品,并显示在页面中。第 64～111 行对应用户单击记录中的"添加购物车"按钮事件。首先需要判断用户是否登录,如果未登录,则跳转到登录页面;如果用户已经登录,则查询数据库 shoppingcart 集合中有没有该用户提交的该商品的记录。如果存在该记录,则修改记录中 num 字段的值;如果不存在该记录,则新增加一条记录。其中,第 83 行判断是否存在该记录,第 84～86 行对应新增一条记录,第 96～100 行对应修改记录中 num 字段的值,因为修改或者新增了购物车中的记录,因此需要修改 tabBar 中购物车右上角的商品数量。

11.2 商品分类页面

商品分类页面展示的内容和项目主页是一样的，只不过展示的形式不一样。用微信开发者工具打开 ColorUI 中的 demo 文件，在 tabBar 中选择"扩展"选项，然后单击"垂直导航"图标，进入"Tab 索引"页面（代码见 pages/plugin/verticalnav），ColorUI 中垂直导航样式如图 11-6 所示。

图 11-6　ColorUI 中垂直导航样式

商品分类页面 classification.wxml 的代码如下：

```
1   <view class='cu-custom'>
2     <view class="cu-bar fixed bg-shadeTop" 
3   style="height:{{CustomBar}}px;padding-top:{{StatusBar}}px;">
4       <view class='content' style='top:{{StatusBar}}px;'>商品分类</view>
5     </view>
6   </view>
7   <swiper class="screen-swiper round-dot" indicator-dots="true" circular="true" autoplay="true"
8   interval="5000" duration="500">
9     <swiper-item wx:for="{{4}}" wx:key="index">
10      <image src="https://image.weilanwl.com/img/4x3-{{index+1}}.jpg" mode='aspectFill'></image>
11    </swiper-item>
12  </swiper>
13
14  <view class="VerticalBox">
```

```
15    <scroll-view class="VerticalNav nav" scroll-y scroll-with-animation scroll-top="{{VerticalNavTop}}"
16  style="height:calc(100vh - 375rpx)">
17      <view class="cu-item {{index==TabCur?'text-green cur':''}}" wx:for="{{list}}" wx:key="index"
18  bindtap='tabSelect' data-id="{{index}}">
19          {{item.name}}
20      </view>
21    </scroll-view>
22
23    <scroll-view class="VerticalMain" scroll-y scroll-with-animation style="height:calc
24  (100vh - 375rpx)" scroll-into-view="main-{{MainCur}}" bindscroll="VerticalMain">
25      <view style="padding-left:10rpx;" wx:for="{{list}}" wx:key="index" id="main-{{index}}"
26  wx:for-index="index">
27        <view class='cu-bar solid-bottom bg-white'>
28          <view class='action'>
29            <text class='cuIcon-title text-green'></text>{{item.name}}</view>
30        </view>
31        <view class="cu-list">
32          <block wx:for="{{contactAll[index]}}" wx:key="id" wx:for-item="gooddetail"
33  wx:for-index="idx">
34            <view class="flex">
35              <view class="flex-twice" bindtap="gogooddetail" data-item="{{contactAll[index][idx]}}">
36                <image style="width:180rpx;height:180rpx"
37  src="{{'cloud://test-0pmu0.7465-test-0pmu0-1300559272/images/'+gooddetail.url[1] }}"
38  mode="aspectFill"></image>
39              </view>
40              <view class="flex-twice padding-sm">
41                <view class="text-lg"><text>{{gooddetail.name}}</text></view>
42                <view class="text-red text-lg">? {{gooddetail.saleprice}}</view>
43              </view>
44              <view class="flex-treble flex align-center justify-end padding-sm" >
45                <view wx:if="{{gooddetail.num > 0}}" ><van-stepper value="{{ 1 }}"
46  input-width="18px" button-size="18px" min="0" max="1000" data-id1="{{index}}" data-id2="{{idx}}"
47  value="{{ gooddetail.num }}" bind:change="onChangeNum" bind:plus="onPlusNum"
48  bind:minus="onMinusNum"/></view>
49              <view wx:else bindtap="onAddCart" data-id1="{{index}}" data-id2="{{idx}}"><image
50  style="width:60rpx;height:60rpx" src="/images/cart.png" mode="aspectFill" ></image></view>
51            </view>
52          </view>
53          <view class="hr"></view>
54        </block>
55      </view>
56    </view>
57  </scroll-view>
58 </view>
```

商品分类页面效果如图11-7所示。其中，代码第7～12行对应页面最上面滑块视图容器；第23～58行对应垂直导航栏；第32～54行显示每条商品的记录，每条商品的记录样式采用ColorUI中Flex比例布局（代码见ColorUI项目中pages/basics/layout），当用户还未添加该商品到购物车时,该记录最右侧显示购物车图标,如果用户已经添加该商品到购物车,则该记录最右侧使用 Vant Weapp 中的 Stepper 步进器显示该商品已添加到购物车的数量。

图 11-7　商品分类页面效果

因为商品分类页面使用了 ColorUI 中的垂直导航样式,所以需要在 classification.wxss 中添加样式,代码如下:

```
1   .VerticalNav.nav {
2     width: 200rpx;
3     white-space: initial;
4   }
5
6   .VerticalNav.nav .cu-item {
7     width: 100%;
8     text-align: center;
9     background-color: #fff;
10    margin: 0;
11    border: none;
12    height: 50px;
13    position: relative;
14  }
15
16  .VerticalNav.nav .cu-item.cur {
17    background-color: #f1f1f1;
18  }
```

```css
19
20  .VerticalNav.nav .cu-item.cur::after {
21    content:"";
22    width: 8rpx;
23    height: 30rpx;
24    border-radius: 10rpx 0 0 10rpx;
25    position: absolute;
26    background-color: currentColor;
27    top: 0;
28    right: 0rpx;
29    bottom: 0;
30    margin: auto;
31  }
32
33  .VerticalBox {
34    display: flex;
35  }
36
37  .VerticalMain {
38    background-color: rgb(255, 255, 255);
39  }
40
41  .hr {
42    height: 2rpx;
43    width: 100%;
44    background: white;
45    border-top: black solid 1px;
46    border-bottom: black solid 1px;
47    opacity: 0.1;
48  }
```

classification.wxss 中的样式和 ColorUI 中垂直导航样式一致，这里不再做详细的介绍。

同样商品分类页面使用了 Vant Weapp 中的 Stepper 步进器，需要在 classification.json 中添加样式，代码如下：

```json
1  {
2    "usingComponents": {
3      "van-stepper": "../../dist/stepper/index"
4    }
5  }
```

相应的 classification.js 代码如下：

```js
1  const app = getApp()
2  const db = wx.cloud.database()
3  Page({
4    data: {
```

```
5        StatusBar: app.globalData.StatusBar,
6        CustomBar: app.globalData.CustomBar,
7        Custom: app.globalData.Custom,
8        TabCur: 0,
9        MainCur: 0,
10       VerticalNavTop: 0,
11       list: [],
12       contactAll: [],
13       load: true,
14       isonload: false
15     },
16     onLoad: function(options) {
17       wx.showLoading({
18         title: '加载中...',
19         mask: true
20       });
21       db.collection('classification').get().then(res =>{
22         let list = [{}];
23         list = res.data
24         var contactAll = []
25         for (let i = 0; i < list.length; i++) {
26           contactAll[i] = new Array()
27         }
28         const MAX_LIMIT = 20
29         db.collection('goods').count().then(res =>{
30           const total = res.total
31           const batchTimes = Math.ceil(total / 20)
32           const tasks = []
33           for (let i = 0; i < batchTimes; i++) {
34             const promise = db.collection('goods').skip(i * MAX_LIMIT).limit(MAX_LIMIT).get();
35             tasks.push(promise)
36           }
37           Promise.all(tasks).then((result) =>{
38             let concatData = []
39             for (let i = 0; i < result.length; i++) {
40               concatData = concatData.concat(result[i].data)
41             }
42             for (let i = 0; i < concatData.length; i++) {
43               let index = concatData[i].classification
44               concatData[i].num = 0
45               contactAll[index].push(concatData[i])
46             }
47             var userInfo = wx.getStorageSync('userInfo');
48             if (userInfo != "") {
49               db.collection('shoppingcart').where({
50                 _openid: userInfo.openid
51               }).get().then(res =>{
52                 var shoppingdetail = res.data
53                 for (let i = 0; i < shoppingdetail.length; i++) {
```

```
54              var idx1 = shoppingdetail[i].classification
55              var item = contactAll[idx1]
56              for (let j = 0; j < item.length; j++) {
57                if (item[j].code == shoppingdetail[i].code) {
58                  contactAll[idx1][j].num = shoppingdetail[i].num
59                }
60              }
61            }
62            this.setData({
63              list: list,
64              listCur: list[0],
65              contactAll: contactAll,
66              isonload: true
67            })
68            wx.hideLoading()
69          })
70        } else {
71          this.setData({
72            list: list,
73            listCur: list[0],
74            contactAll: contactAll,
75            isonload: true
76          })
77          wx.hideLoading()
78        }
79      })
80    })
81  })
82  },
83
84  onShow: function() {
85    var isonload = this.data.isonload
86    if (isonload) {
87      var userInfo = wx.getStorageSync('userInfo');
88      var contactAll = this.data.contactAll
89      if (userInfo != "") {
90        for (let i = 0; i < contactAll.length; i++) {
91          for (let j = 0; j < contactAll[i].length; j++) {
92            contactAll[i][j].num = 0
93          }
94        }
95        db.collection('shoppingcart').where({
96          _openid: userInfo.openid
97        }).get().then(res => {
98          var shoppingdetail = res.data
99          for (let i = 0; i < shoppingdetail.length; i++) {
100           var idx1 = shoppingdetail[i].classification
101           var item = contactAll[idx1]
102           for (let j = 0; j < item.length; j++) {
```

```javascript
103              if (item[j].code == shoppingdetail[i].code) {
104                contactAll[idx1][j].num = shoppingdetail[i].num
105              }
106            }
107          }
108          this.setData({
109            contactAll: contactAll,
110          })
111        })
112      }
113    }
114  },
115
116  gogooddetail: function(event) {
117    var goodinfo = event.currentTarget.dataset.item
118    wx.navigateTo({
119      url: '../gooddetail/gooddetail?goodinfo=' + JSON.stringify(goodinfo)
120    })
121  },
122
123  onAddCart: function(event) {
124    var userInfo = wx.getStorageSync('userInfo');
125    if (userInfo == "") {
126      wx.navigateTo({
127        url: '../auth/auth'
128      })
129    } else {
130      var id1 = event.currentTarget.dataset.id1
131      var id2 = event.currentTarget.dataset.id2
132      var contactAll = this.data.contactAll
133      contactAll[id1][id2].num = 1
134      this.setData({
135        contactAll: contactAll
136      })
137      var goodinfo = contactAll[id1][id2]
138      goodinfo.submitdate = db.serverDate()
139      goodinfo.openid = userInfo.openid
140      delete goodinfo._id
141      db.collection('shoppingcart').add({
142        data: goodinfo,
143      })
144        .then(res =>{
145          var sum = app.globalData.goodnum + 1
146          app.globalData.goodnum = sum
147          wx.setTabBarBadge({
148            index: 2,
149            text: sum + ""
150          })
151        })
```

```js
            .catch(console.error)
    }
},

onPlusNum: function(event) {
    var sum = app.globalData.goodnum + 1
    app.globalData.goodnum = sum
    wx.setTabBarBadge({
        index: 2,
        text: sum + ""
    })
},

onMinusNum: function(event) {
    var sum = app.globalData.goodnum - 1
    app.globalData.goodnum = sum
    wx.setTabBarBadge({
        index: 2,
        text: sum + ""
    })
},

onChangeNum: function(event) {
    var id1 = event.currentTarget.dataset.id1
    var id2 = event.currentTarget.dataset.id2
    var contactAll = this.data.contactAll
    contactAll[id1][id2].num = event.detail
    var code = contactAll[id1][id2].code
    var userInfo = wx.getStorageSync('userInfo');
    this.setData({
        contactAll: contactAll
    })
    db.collection('shoppingcart').where({
        code: code,
        _openid: userInfo._openid
    }).get().then(res =>{
        if (event.detail == 0) {
            db.collection('shoppingcart').doc(res.data[0]._id).remove()
                .then(console.log)
                .catch(console.error)
        } else {
            db.collection('shoppingcart').doc(res.data[0]._id).update({
                data: {
                    num: event.detail
                }
            })
                .then(console.log)
                .catch(console.error)
        }
```

```
201        })
202      },
203
204      tabSelect(e) {
205        this.setData({
206          TabCur: e.currentTarget.dataset.id,
207          MainCur: e.currentTarget.dataset.id,
208          VerticalNavTop: (e.currentTarget.dataset.id - 1) * 50
209        })
210      },
211
212      VerticalMain(e) {
213        let that = this;
214        let list = this.data.list;
215        let tabHeight = 0;
216        if (this.data.load) {
217          for (let i = 0; i < list.length; i++) {
218            let view = wx.createSelectorQuery().select("#main-" + list[i].id);
219            view.fields({
220              size: true
221            }, data =>{
222              list[i].top = tabHeight;
223              tabHeight = tabHeight + data.height;
224              list[i].bottom = tabHeight;
225            }).exec();
226          }
227          that.setData({
228            load: false,
229            list: list
230          })
231        }
232        let scrollTop = e.detail.scrollTop + 20;
233        for (let i = 0; i < list.length; i++) {
234          if (scrollTop > list[i].top && scrollTop < list[i].bottom) {
235            that.setData({
236              VerticalNavTop: (list[i].id - 1) * 50,
237              TabCur: list[i].id
238            })
239            return false
240          }
241        }
242      }
243    })
```

代码第16~82行对应页面加载事件,其中第21行查询classification数据库集合中商品分类信息,第24~80行查询商品记录,因为微信小程序端最多一次取20条记录,所以很可能一个请求无法取出所有数据,需要分批次取。第37~46行在Promise.all中把所有批次取到的数据进行汇总,存放到concatData数组中,因为在页面中展示的数据需要按照商品的分类提供一个二维数组,所以代码第42~46行把获取到的商品记录按商品分类转换为二维数组 contactAll，二维数组 contactAll 数据形式如图 11-8 所示。为了保证 shoppingcart 数据库集合中的记录能和该页面中的 Stepper 步进器显示的数量一致，代码第 49～61 行需要查询该用户在 shoppingcart 数据库集合中的记录，并把购物车中商品的数量同步到更新到 Stepper 步进器。当用户在其他页面，比如在项目主页中添加商品到购物车或者在购物车页面中对商品进行修改或者删除操作，再回到商品分类页面，根据表 2-5，这时 onLoad 事件将不再被调用，而其他页面的操作会影响商品分类页面 Stepper 步进器的值，因此需要在 onShow 事件中更新 contactAll 中记录 num 字段的值。这里需要说明的是，虽然首次打开商品分类页面先调用 onLoad 事件，再调用 onShow 事件，但是 onLoad 也是异步执行的，所以有可能 onLoad 事件还未执行完毕，onShow 事件就被调用，而 onShow 事件只更新 contactAll 中记录 num 字段的值，data 变量中设置了 isonload 变量，作用是首次加载商品分类页面只执行 onLoad 事件，不执行 onShow 事件。代码第 95～111 行查询 shoppingcart 数据库集合，把购物车中商品的数量更新到商品分类页面中的 Stepper 步进器中。代码第 116～121 行对应用户单击商品图片，页面跳转到商品详情页面。代码第 123～154 行对应用户单击购物车图标，把商品记录添加到购物车中。代码第 156～172 行对应用户单击 Stepper 步进器中的"－"和"＋"按钮，更新全局变量 app.globalData.goodnum 的值，并更新 tabBar 中购物车右上角的数字。代码第 174～202 行 onChangeNum 对应 Stepper 步进器当绑定值变化时触发的事件，步进器的值为 event.detail，当 event.detail 值为 0 时，表示要从购物车中删除该记录，其他情况则只需要更新记录中 num 字段的值即可，其中代码第 189～191 行对应从购物车中删除该记录，代码第 193～199 行对应更新该记录中 num 字段的值。代码第 204～242 行对应 ColorUI 中提供的代码，tabSelect 对应用户单击左侧商品分类事件，VerticalMain 对应滑动滚动条事件。

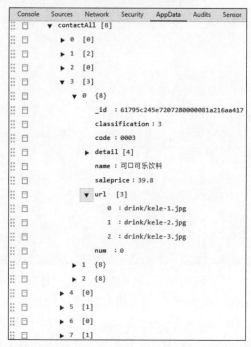

图 11-8　AppData 中商品记录中二维数组 contactAll 数据形式

11.3 商品详情页面

用户在商品分类页面中单击商品图片,进入商品详情页面,商品详情页面使用了 Vant Weapp 中 GoodsAction 商品导航组件,如图 10-7 所示,本项目实现了"购物车"和"加入购物车"按钮的功能。

商品详情页 gooddetail.wxml 实现代码如下:

```
1   <cu-custom bgColor = "bg-gradual-pink" isBack = "{{true}}">
2     <view slot = "backText">返回</view>
3     <view slot = "content">商品详情</view>
4   </cu-custom>
5
6   <swiper class = "screen-swiper round-dot}}" indicator-dots = "true" circular = "true" autoplay = "true"
7   interval = "5000" duration = "500">
8     <swiper-item wx:for = "{{goodinfo.url}}" wx:key = "index">
9       <image src = "{{'cloud://test-0pmu0.7465-test-0pmu0-1300559272/images/' + item}}"
10  mode = "aspectFill"></image>
11    </swiper-item>
12  </swiper>
13  <view class = "padding radius shadow bg-red">
14    <text class = "text-lg">¥{{goodinfo.saleprice}}/份</text>
15  </view>
16  <view class = "padding bg-white">
17    <text class = 'cuIcon-title text-green'></text>
18    <text class = "text-lg">商品详情</text>
19  </view>
20  <view wx:for = "{{goodinfo.detail}}" wx:key = "index">
21    <image style = "width:100%"
22  src = "{{'cloud://test-0pmu0.7465-test-0pmu0-1300559272/images/' + item}}" mode = "widthFix"></image>
23  </view>
24  <van-goods-action>
25    <van-goods-action-icon icon = "wap-home-o" text = "主页" bind:click = "onClickIcon" />
26    <van-goods-action-icon icon = "cart-o" info = "{{goodcount}}" text = "购物车"
27  bind:click = "onClickShoppingCart" />
28    <van-goods-action-button text = "加入购物车" type = "warning" bind:click = "onClickAddgood" />
29    <van-goods-action-button text = "立即购买" bind:click = "onClickButton" />
30  </van-goods-action>
```

商品详情页面显示效果如图 11-9 所示。代码第 6~12 行对应页面最上面滑块视图容器,显示商品介绍图片。代码第 13~23 行对应商品详情图片。代码第 24~30 行对应 Vant Weapp 中 GoodsAction 商品导航组件,因此需要在 gooddetail.json 文件中引入相应组件:

```
1   {
2     "usingComponents": {
3       "van-goods-action": "../../dist/goods-action",
4       "van-goods-action-icon": "../../dist/goods-action-icon",
```

```
5        "van-goods-action-button": "../../dist/goods-action-button"
6    }
7 }
```

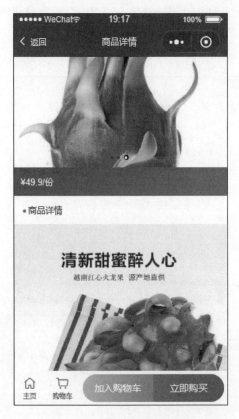

图 11-9　商品详情页面效果

相应的 gooddetail.js 代码如下：

```
1 Page({
2   data: {
3     goodinfo:{}
4   },
5   onLoad: function (options) {
6     const goodinfo = JSON.parse(options.goodinfo)
7     this.setData({ goodinfo: goodinfo})
8   },
9 })
```

商品详情页面的逻辑相对比较简单，从上一页面传递的参数中提取包含商品信息的字符串，通过 JSON.parse() 把字符串转换为 JSON 对象，从而获得该商品的数据存入 data 中的 goodinfo 变量。

11.4 购物车页面

购物车页面和第 10 章网上书城微信小程序购物车页面相似。购物车页面使用了 Vant Weapp 中 Card 商品卡片和 SubmitBar 提交订单栏组件，Card 商品卡片样式如图 10-4 所示，SubmitBar 提交订单栏如图 10-9 所示。

购物车页面 shoppingcart.wxml 实现代码如下：

```
1   <cu-custom bgColor="bg-gradual-pink" isBack="{{false}}">
2     <view slot="backText">返回</view>
3     <view slot="content">购物车</view>
4   </cu-custom>
5
6   <view wx:for="{{shoppingcart_list}}" wx:key="index">
7     <van-card data-id="{{item._id}}" price="售价:{{item.saleprice}}元"
8   thumb="{{'cloud://test-0pmu0.7465-test-0pmu0-1300559272/images/' + item.url[0]}}"
9   title="商品名称:{{item.name}}">
10      <view slot="bottom">
11        <van-col span="4" offset="20">
12          <van-checkbox value="{{ item.checked }}" data-id="{{item._id}}" data-index="{{index}}"
13  checked-color="#ff3366" bind:change="onSelectGood">
14          </van-checkbox>
15        </van-col>
16      </view>
17      <view slot="tags">
18        <van-stepper value="{{ 1 }}" button-size="20px" min="0" max="1000" value="{{ item.num }}"
19  data-id="{{item._id}}" data-index="{{index}}" bind:change="onChangeNum" bind:plus="onPlusNum"
20  bind:minus="onMinusNum" />
21      </view>
22    </van-card>
23    <view class="container">
24      <view class="divLine"></view>
25    </view>
26  </view>
27
28  <van-submit-bar price="{{ totalprice }}" button-text="去结算" bind:submit="onClickButton"
29  tip="{{ true }}">
30    <van-tag type="primary" style="margin-left:20px">
31      <van-checkbox value="{{ checked }}" checked-color="#ff3366" bind:change="onSelectAll">
32      </van-checkbox>
33    </van-tag>
34    <view slot="tip">
35      目前商城不支持快递配送,只支持到店取货
36    </view>
37  </van-submit-bar>
38  <van-dialog id="van-dialog" />
```

购物车页面效果如图 11-10 所示,在页面上使用了 Card 商品卡片和 SubmitBar 提交订单

栏组件；代码第 6~26 行对应 Card 商品卡片组件，其中代码第 12 行使用了 Vant Weapp 中 Checkbox 复选框，代码第 18~20 行使用了 Vant Weapp 中 Stepper 步进器；代码第 28~37 行对应 SubmitBar 提交订单栏组件，代码第 38 行对应 Vant Weapp 中的对话框组件。

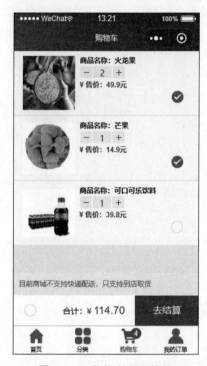

图 11-10　购物车页面效果

购物车页面使用了 Vant Weapp 中多个组件，因此需要在 shoppingcart.json 文件中引入组件：

```
1  {
2    "usingComponents": {
3      "van-card": "../../dist/card",
4      "van-stepper": "../../dist/stepper",
5      "van-col": "../../dist/col",
6      "van-checkbox": "../../dist/checkbox",
7      "van-submit-bar": "../../dist/submit-bar",
8      "van-dialog": "../../dist/dialog"
9    }
10 }
```

页面中引入了分割线，页面底端使用的 SubmitBar 提交订单栏组件也需要设置页面 padding-bottom 值，shoppingcart.wxss 页面样式如下：

```
1  page {
2    padding-bottom: 200rpx;
3  }
4  .divLine{
```

```
5    border-left: 1rpx dashed #a7a8a8;
6    width: 100%;
7    height: 5rpx;
8  }
```

相应的 shoppingcart.js 代码如下:

```
1   const app = getApp()
2   const db = wx.cloud.database()
3   import Dialog from '../../dist/dialog/dialog';
4   Page({
5     data: {
6       shoppingcart_list: {},
7       checked: false,
8       totalprice: 0,
9     },
10  
11    caculateTotalPrice() {
12      var totalprice = 0
13      var shoppingcart_list = this.data.shoppingcart_list
14      for (var i = 0; i < shoppingcart_list.length; i++) {
15        if (shoppingcart_list[i].checked)
16          totalprice = totalprice + shoppingcart_list[i].saleprice * shoppingcart_list[i].num
17      }
18      this.setData({
19        totalprice: parseInt(totalprice * 100)
20      })
21    },
22  
23    checkSelect() {
24      var shoppingcart_list = this.data.shoppingcart_list
25      var isAllSelect = true
26      for (var i = 0; i < shoppingcart_list.length; i++) {
27        if (! shoppingcart_list[i].checked && isAllSelect)
28          isAllSelect = false
29      }
30      this.setData({
31        checked: isAllSelect
32      })
33    },
34  
35    onShow: function(options) {
36      var userInfo = wx.getStorageSync('userInfo');
37      if (userInfo == "") {
38        wx.navigateTo({
39          url: '../auth/auth'
40        })
41      } else {
42        db.collection('shoppingcart').where({
```

```js
                _openid: userInfo._openid
         }).get().then(res =>{
            this.setData({
               shoppingcart_list: res.data,
            })
            this.caculateTotalPrice()
            this.checkSelect()
         })
      }
   },

   onSelectGood(event) {
      var index = event.currentTarget.dataset.index
      var id = event.currentTarget.dataset.id
      var shoppingcart_list = this.data.shoppingcart_list
      shoppingcart_list[index].checked = event.detail
      this.setData({
         shoppingcart_list: shoppingcart_list
      })
      this.caculateTotalPrice();
      this.checkSelect()
      db.collection('shoppingcart').doc(id).update({
         data: {
            checked: event.detail
         },
         success: function(res) {
            console.log(res)
         }
      })
   },

   onPlusNum: function(event) {
      var sum = app.globalData.goodnum + 1
      app.globalData.goodnum = sum
      wx.setTabBarBadge({
         index: 2,
         text: sum + ""
      })
   },

   onMinusNum: function(event) {
      var sum = app.globalData.goodnum - 1
      app.globalData.goodnum = sum
      wx.setTabBarBadge({
         index: 2,
         text: sum + ""
      })
   },
```

```
92    onChangeNum(event) {
93      var index = event.currentTarget.dataset.index
94      var id = event.currentTarget.dataset.id
95      var shoppingcart_list = this.data.shoppingcart_list
96      if (event.detail == 0) {
97        shoppingcart_list[index].num = 1
98        this.setData({
99          shoppingcart_list: shoppingcart_list
100       })
101       Dialog.confirm({
102         message: '确认要删除宝贝吗?'
103       }).then(() =>{
104         shoppingcart_list.splice(index, 1);
105         this.setData({
106           shoppingcart_list: shoppingcart_list
107         })
108         db.collection('shoppingcart').doc(id).remove({
109           success: function(res) {
110             wx.showToast({
111               title: '删除成功...',
112               icon: 'none',
113               duration: 1500
114             })
115           },
116           fail: console.error
117         })
118         Dialog.close();
119       }).catch(() =>{
120         var sum = app.globalData.goodnum + 1
121         app.globalData.goodnum = sum
122         wx.setTabBarBadge({
123           index: 2,
124           text: sum + ""
125         })
126         Dialog.close();
127       });
128     } else {
129       shoppingcart_list[index].num = event.detail
130       if (!shoppingcart_list[index].checked) {
131         db.collection('shoppingcart').doc(id).update({
132           data: {
133             checked: true
134           },
135           success: function(res) {}
136         })
137       }
138       shoppingcart_list[index].checked = true
139       this.setData({
140         shoppingcart_list: shoppingcart_list
```

```
141         })
142         this.caculateTotalPrice();
143         this.checkSelect()
144         db.collection('shoppingcart').doc(id).update({
145           data: {
146             num: event.detail
147           },
148           success: function(res) {
149             console.log(res)
150           }
151         })
152       }
153     },
154
155     onSelectAll(event) {
156       var shoppingcart_list = this.data.shoppingcart_list
157       var checked = this.data.checked
158       for (var i = 0; i < shoppingcart_list.length; i++) {
159         shoppingcart_list[i].checked = !checked
160         db.collection('shoppingcart').doc(shoppingcart_list[i]._id).update({
161           data: {
162             checked: !checked
163           },
164           success: function(res) {}
165         })
166       }
167       this.setData({
168         shoppingcart_list: shoppingcart_list,
169         checked: !checked
170       })
171       this.caculateTotalPrice();
172     },
173
174     onClickButton() {
175       var shoppingcart_list = this.data.shoppingcart_list
176       var list = []
177       for (var i = 0; i < shoppingcart_list.length; i++) {
178         if (shoppingcart_list[i].checked)
179           list.push(shoppingcart_list[i]);
180       }
181       wx.navigateTo({
182         url: '../order/order?orderinfo=' + JSON.stringify(list)
183       })
184     }
185   })
```

代码第 11～21 行计算选中订单的总价，选中商品的价格为：数量×售价，订单的总价 totalprice 计算所有选中商品价格的总和，单位为"分"；shoppingcart_list[i].checked 对应商品是否被选中，如果被选中，则该值为 true，反之为 false。shoppingcart_list[i].checked

对应图 11-10 中每件商品右侧的 checkbox 的状态。代码第 23～33 行判断 SubmitBar 提交订单栏组件中的 checkbox 状态是否为全选，如果每件商品后面的 checkbox 都是选中状态，那么 SubmitBar 提交订单栏组件中的 checkbox 状态也应该是选中状态。代码第 35～52 行对应页面 onShow 事件，查询数据库 shoppingcart 集合中_openid 字段为用户_openid 的记录。代码第 54～72 行对应图 11-10 中每件商品后的 checkbox 单击事件，单击 checkbox 需要调用 caculateTotalPrice 重新计算订单总价，调用 checkSelect 判断是否全选，把该记录是否被选中的状态写入数据库。代码第 74～90 对应 Stepper 步进器"－"按钮和"＋"按钮单击事件，需要在这 2 个事件中更新 tabBar 中购物车右上角显示的数字。代码第 92～153 行对应图 11-10 中 Stepper 步进器当绑定值变化时触发的事件，在该事件中需要把购买商品的数量写入数据库，如果该记录的 checkbox 未被选中，则需要设置该记录为选中状态，代码第 96～128 行对应商品的数量减少到 0 时询问用户是否把该商品从购物车中删除，代码第 101～103 行调用 Vant Weapp 对话框组件询问用户是否删除该商品，如果用户单击"确认"按钮，则把该商品记录从数据库 shoppingcart 集合中删除，并从 data 中的 shoppingcart_list 中删除该记录。代码第 155～172 行对应单击 SubmitBar 提交订单栏组件中的 checkbox 事件，使所有记录处于"全选"或者"全不选"的状态。代码第 174～184 行把选中的记录转换为字符串，作为参数传递到提交订单页面。

11.5 提交订单页面

提交订单页面使用了 Vant Weapp 中的 SubmitBar 提交订单栏组件，使用方法和购物车页面的方法相同；提交订单页面 order.wxml 的实现代码如下：

```
1   <cu-custom bgColor="bg-gradual-pink" isBack="{{true}}">
2     <view slot="backText">返回</view>
3     <view slot="content">填写订单</view>
4   </cu-custom>
5
6   <view class="cu-form-group">
7     <view class="title">支付方式</view>
8     <view class="title">现场交易
9       <text class="cuIcon-locationfill text-orange"></text>
10    </view>
11  </view>
12
13  <view class="bg-white padding margin-top">
14    <view class="grid col-4 grid-square">
15      <view class="bg-img" wx:for="{{orderinfo.length<3?orderinfo.length:3}}" wx:key="index">
16        <image style="width:180rpx;height:180rpx"
17  src="{{'cloud://test-0pmu0.7465-test-0pmu0-1300559272/images/'+orderinfo[index].url[0]}}"
18  mode="aspectFill"></image>
19      </view>
20      <view class="text-center">
21        <text>共{{orderinfo.length}}件...</text>
```

```
22        </view>
23      </view>
24    </view>
25
26    <view class = "cu-form-group margin-top">
27      <view class = "title">商品金额</view>
28      <view class = "title">￥{{totalprice}}
29    </view>
30  </view>
31  <view class = "cu-form-group">
32    <view class = "title">运费</view>
33    <view class = "title">
34      <text style = "color:red">+￥0.00</text>
35    </view>
36  </view>
37  <view class = "cu-form-group">
38    <view class = "title">立减</view>
39    <view class = "title">
40      <text style = "color:red">-￥0.00</text>
41    </view>
42  </view>
43
44  <van-submit-bar price = "{{ totalprice * 100}}" button-text = "提交订单" bind:submit = "onClickButton">
45  </van-submit-bar>
```

提交订单页面效果如图 11-11 所示，订单商品封面展示采用了 ColorUI 中的 Grid 布局，最多显示 3 种商品的封面。和购物车页面一行，需要在 order.json 中引入 SubmitBar 提交订单栏组件：

图 11-11　提交订单页面效果

```
1  {
2    "usingComponents": {
3      "van-submit-bar": "../../dist/submit-bar"
4    }
5  }
```

相应的 order.js 代码如下：

```
1   const app = getApp()
2   const db = wx.cloud.database()
3   Page({
4     data: {
5       orderinfo: [],
6       totalprice: 0
7     },
8
9     onLoad: function(options) {
10      var orderinfo = JSON.parse(options.orderinfo)
11      var totalprice = 0
12      for (var i = 0; i < orderinfo.length; i++) {
13        totalprice = totalprice + orderinfo[i].saleprice * orderinfo[i].num
14      }
15      totalprice = parseInt(totalprice * 100) / 100
16      totalprice = totalprice.toFixed(2)
17      this.setData({
18        orderinfo: orderinfo,
19        totalprice: totalprice
20      })
21    },
22
23    onClickButton: function(event) {
24      var orderinfo = this.data.orderinfo
25      var totalprice = this.data.totalprice
26      var userInfo = wx.getStorageSync('userInfo');
27      db.collection('order').add({
28        data: {
29          orderinfo: orderinfo,
30          totalprice: totalprice,
31          submitorder: db.serverDate(),
32          _openid: userInfo._openid
33        }
34      })
35      .then(res =>{
36        var sum = app.globalData.goodnum
37        for (var i = 0; i < orderinfo.length; i++) {
38          db.collection('shoppingcart').doc(orderinfo[i]._id).remove()
39          sum = sum - orderinfo[i].num
40        }
41        app.globalData.goodnum = sum
```

```
42              console.log(sum)
43              if (sum == 0) {
44                wx.removeTabBarBadge({
45                  index: 2
46                })
47              } else {
48                wx.setTabBarBadge({
49                  index: 2,
50                  text: sum + ""
51                })
52              }
53              wx.showToast({
54                title: '订单提交成功...',
55                icon: 'none',
56                duration: 1500
57              })
58              wx.switchTab({
59                url: '../home/home'
60              })
61            })
62            .catch(console.error)
63        }
64   })
```

代码第 9~21 行在页面加载事件中计算订单总价，精确到小数点后 2 位。代码第 23~63 行对应单击"提交订单"按钮事件，把订单数据加入数据库 order 集合，并把该记录从购物车集合 shoppingcart 中删除，同时更新 tabBar 中购物车右上角的数字，订单提交成功后页面跳转到项目主页。

11.6 我的订单页面

我的订单页面 myorder.wxml 的实现代码如下：

```
1   <cu-custom bgColor = "bg-gradual-pink" isBack = "{{true}}">
2     <view slot = "backText">返回</view>
3     <view slot = "content">我的订单</view>
4   </cu-custom>
5
6   <view wx:for = "{{myorder_list}}" wx:for-item = "myorder" wx:key = "index">
7     <view class = "cu-bar bg-white margin-top solid-bottom">
8       <view class = "action">
9         <text class = "cuIcon-title text-blue"></text>购买商品
10      </view>
11      <view class = "action"></view>
12    </view>
13    <view class = "bg-white padding">
14      <view class = "grid col-4 grid-square">
```

```
15        <view class = "bg - img" wx:for = "{{myorder.orderinfo.length<3? myorder.orderinfo.length:3}}"
16  wx:key = "index">
17          <image style = "width:180rpx;height:180rpx"
18  src = "{{'cloud://test-0pmu0.7465-test-0pmu0-1300559272/images/'+myorder.orderinfo[index].url[0]}}"
19  mode = "aspectFill"></image>
20        </view>
21        <view class = "text - center">
22          <view class = "text - lg" style = "color:red">? {{myorder.totalprice}}</view>
23          <view>共{{myorder.orderinfo.length}}件</view>
24        </view>
25      </view>
26    </view>
27  </view>
```

我的订单页面中每个订单的显示样式采用和提交订单页面中订单商品封面展示相同的样式。我的订单页面效果如图11-12所示。

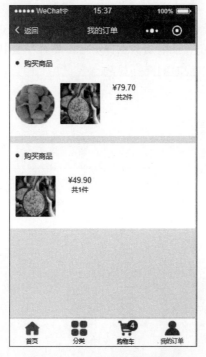

图 11-12 我的订单页面效果

相应的myorder.js代码如下：

```
1  const db = wx.cloud.database()
2  Page({
3    data:{
4      myorder_list:[]
5    },
6
```

```
7    onShow: function () {
8      var userInfo = wx.getStorageSync('userInfo');
9      if (userInfo == "") {
10       wx.navigateTo({
11         url: '../auth/auth'
12       })
13     } else {
14       db.collection('order').where({
15         _openid: userInfo._openid
16       }).orderBy('submitorder', 'desc').get().then(res =>{
17         this.setData({
18           myorder_list: res.data
19         })
20       })
21     }
22   },
23 })
```

代码第7~22行在页面onShow事件中，先判断缓冲中是否有用户信息，如果没有则跳转到登录页面，如果用户已经登录则从数据库order集合中查找_openid字段等于本人的订单记录，并按照用户提交订单的时间降序排列。

第 12 章

会议室预约微信小程序

会议室预约微信小程序包含 5 个页面：

```
1  "pages": [
2    "pages/home/home",              //项目主页
3    "pages/room/room",              //会议室预约情况页面
4    "pages/meetingform/meetingform", //填写预约单页面
5    "pages/login/login",            //登录页面
6    "pages/myorder/myorder"         //我的预约页面
7  ],
```

其中，包含 4 个数据库集合：Verify（存储短信验证码等登录信息）、room（会议室信息）、order（预约单）、contact（允许预约会议室人员的信息）。权限均为"所有用户可读，仅创建者可读写"。要运行本项目程序需要先在数据库中创建这 4 个集合。

本项目案例使用的登录页面和第 9 章报修类微信小程序使用的登录页面一样，在此不再重复说明。登录页面效果如图 12-1 所示。

12.1 项目主页

项目主页提供了会议室的信息，包括房间号、管理员信息和会议室使用规则。项目主页（pages/home/home）样式采用 ColorUI 中的主页样式（代码见 pages/basics/home），ColorUI 中的主页样式如图 12-2 所示。

图 12-1　会议室预约微信小程序
　　　　　登录页面效果

图 12-2　ColorUI 中的主页样式

本案例项目主页 home.wxml 的代码如下：

```
1   <cu-custom bgImage = "/images/sylb.jpg" isBack = "{{false}}">
2     <view slot = "backText">返回</view>
3     <view slot = "content">会议室</view>
4   </cu-custom>
5
6   <view class = 'nav-list margin-top'>
7     <navigator open-type = "navigate" hover-class = 'none'
8   url = "/pages/room/room? roomID = {{item.roomID}}" class = "nav-li bg-orange" wx:for = "{{rooms}}"
9   wx:key = "index">
10      <view class = "nav-title">{{item.roomID}}</view>
11      <view class = "nav-name">管理员:{{item.adminname}}{{item.adminphone_cornet}}</view>
12      <text class = 'cuIcon-brand'></text>
13    </navigator>
14  </view>
15
16  <view class = 'operate'>
17    <button class = "cu-btn bg-green margin-tb-xs lg" bindtap = "meetingRules">使用规则</button>
18    <button class = "cu-btn bg-green margin-tb-xs lg" bindtap = "myOrder">我的预约</button>
19  </view>
```

代码第 6~14 行使用了 ColouUI 中主页样式效果，显示会议室信息，包括房间号、管理员信息等。代码第 16~19 显示 2 个按钮，分别对应"使用规则"和"我的预约"功能。

这里需要在 home.wxss 引入如下样式：

```css
.nav-list {
  display: flex;
  flex-wrap: wrap;
  padding: 0px 30rpx 0px;
  justify-content: space-between;
}
.nav-li {
  padding: 30rpx;
  border-radius: 12rpx;
  width: 45%;
  margin: 0 2.5% 40rpx;
  background-image: url(https://image.weilanwl.com/color2.0/cardBg.png);
  background-size: cover;
  background-position: center;
  position: relative;
  z-index: 1;
}
.nav-li::after {
  content: "";
  position: absolute;
  z-index: -1;
  background-color: inherit;
  width: 100%;
  height: 100%;
  left: 0;
  bottom: -10%;
  border-radius: 10rpx;
  opacity: 0.2;
  transform: scale(0.9, 0.9);
}
.nav-li.cur {
  color: #fff;
  background: rgb(94, 185, 94);
  box-shadow: 4rpx 4rpx 6rpx rgba(94, 185, 94, 0.4);
}
.nav-title {
  font-size: 32rpx;
  font-weight: 300;
}
.nav-title::first-letter {
  font-size: 40rpx;
  margin-right: 4rpx;
}
.nav-name {
  font-size: 28rpx;
  text-transform: Capitalize;
  margin-top: 20rpx;
  position: relative;
}
```

```css
50  .nav-name::before{
51    content:"";
52    position:absolute;
53    display:block;
54    width:40rpx;
55    height:6rpx;
56    background:#fff;
57    bottom:0;
58    right:0;
59    opacity:0.5;
60  }
61  .nav-name::after{
62    content:"";
63    position:absolute;
64    display:block;
65    width:100rpx;
66    height:1px;
67    background:#fff;
68    bottom:0;
69    right:40rpx;
70    opacity:0.3;
71  }
72  .nav-li text{
73    position:absolute;
74    right:30rpx;
75    top:30rpx;
76    font-size:52rpx;
77    width:60rpx;
78    height:60rpx;
79    text-align:center;
80    line-height:60rpx;
81  }
82  .operate{
83    display:flex;
84    flex-wrap:nowrap;
85    justify-content:space-around;
86    width:100%;
87    padding:5rpx 0;
88    background-color:white;
89    position:fixed;
90    bottom:0;
91  }
```

项目主页效果如图 12-3 所示，相应的 home.js 代码如下：

```js
1  const db = wx.cloud.database()
2  Page({
3    data:{
4      rooms:[],
```

图 12-3　项目主页效果

```
5     },
6
7     onLoad: function (options) {
8       db.collection('room').get().then(res =>{
9         this.setData({rooms:res.data})
10      })
11    },
12
13    meetingRules: function (event) {
14      wx.showLoading({
15        title: '文档打开中...',
16      })
17      wx.cloud.downloadFile({
18        fileID: 'cloud://test-0pmu0.7465-test-0pmu0-1300559272/files/计算机与人工智
19  能学院会议室使用规则.docx'
20      }).then(res =>{
21        console.log(res.tempFilePath)
22        const filePath = res.tempFilePath
23        wx.openDocument({
24          filePath: filePath,
25          success: res =>{
26            console.log('打开文档成功')
27            wx.hideLoading()
28          }
29      })
```

```
30          }).catch(error =>{
31            // handle error
32          })
33      },
34
35      myOrder:function(event)
36      {
37        var phone = wx.getStorageSync('phone');
38        if (phone ! = "") {
39          wx.navigateTo({
40            url: '../myorder/myorder'
41          })
42        }
43        else{
44          wx.navigateTo({
45            url: '../login/login'
46          })
47        }
48      }
49  }))
```

代码第 7~11 行查询数据库集合 room 中的所有会议室信息。代码第 13~33 行打开云储存中的会议室使用规则，在这里会议室使用规则可以是一个 Word 或者 PDF 文件，管理员先把这个文件上传到云存储中，把文件的 fileID 写入第 18 行，代码第 17~20 行从云存储中下载文件，代码第 23~29 行打开文件进行浏览，代码第 35~49 行对应单击"我的预约"按钮事件，如果缓存中存在用户手机号，则跳转到我的预约页面，否则跳转到登录页面。想要实现如图 12-13 所示的效果，需要在数据库集合 room 中添加会议室信息，数据库集合 room 字段信息如图 12-4 所示，包括管理员姓名（adminname）、管理员手机号（adminphone）、adminphone_cornet（管理员单位短号）、会议室容纳人数（capacity）、房间号（roomID）和房间照片（url）。

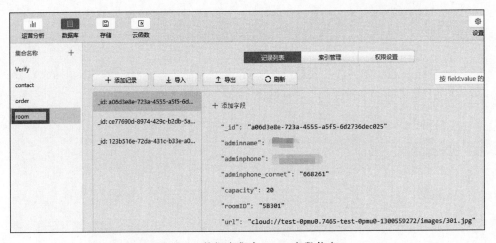

图 12-4　数据库集合 room 字段信息

12.2　会议室预约情况页面

选择样式：用微信开发者工具打开 ColorUI 中的 demo 文件，在 tabBar 中选择"组件"选项，然后单击"卡片"，进入"卡片"页面，选择新闻主页样式如图 6-7 所示。

在 ColorUI 中打开 component/card/card.wxml 页面找到对应代码，编辑后本案例会议室预约情况页面（pages/room/room）的样式 room.wxml 代码如下：

```
1   <cu-custom bgImage="/images/sylb.jpg" isBack="{{true}}">
2     <view slot="backText">返回</view>
3     <view slot="content">会议室预约情况</view>
4   </cu-custom>
5
6   <scroll-view class="scroll-view_H" scroll-x>
7     <view class='list' style='width:{{width}}rpx'>
8       <view bindtap="select" wx:for="{{calendar}}" wx:for-item="item" class='listItem
9   {{index==currentIndex?"current":""}}' wx:key="index" data-date="{{item.date}}" data-
10  index="{{index}}">
11        <text class='name'>{{item.week}}</text>
12        <text class='date'>{{item.date}}</text>
13      </view>
14    </view>
15  </scroll-view>
16
17  <view class="cu-list menu-avatar comment solids-top">
18    <block wx:for="{{records}}" wx:key="index">
19      <view class="cu-item">
20        <view class="cu-avatar round" style="background-image:url({{item.avatarUrl}});"></view>
21        <view class="content">
22          <view class=" flex justify-between">
23            <view class="text-gray text-df">申请人:{{item.name}}({{item.phone_cornet}})</view>
24            <view>
25              <view class="cu-avatar round sm bg-green">{{index+1}}</view>
26            </view>
27          </view>
28          <view class="text-gray text-content text-df">
29            会议时间:{{item.selectDate}} {{item.startTime}}-{{item.endTime}}
30          </view>
31          <view class="text-gray text-content text-df">
32            会议地点:{{item.roomID}}
33          </view>
34          <view class="bg-pink padding-sm radius margin-top-sm text-sm">
35            <view class="flex">
36              <view class="flex-sub">会议主题:{{item.title}}</view>
37            </view>
38          </view>
39          <view class="margin-top-sm flex justify-between">
40            <view class="text-gray text-df">提交时间:{{item.submitdate}}</view>
```

```
41              <view>
42                <text class="cuIcon-deletefill text-red margin-left-sm" data-index="{{index}}"
43  bindtap="deleteRecord"></text>
44              </view>
45            </view>
46          </view>
47        </view>
48        <view class="container">
49          <!-- 分割线 -->
51          <view class="divLine"></view>
52        </view>
53      </block>
54  </view>
55
56  <view class='operate'>
57    <button class="cu-btn bg-green margin-tb-xs lg" bindtap="gotoAppointment">会议室预约</button>
58  </view>
```

代码第 6~15 行对应日期导航栏,用户只能查看和预约最近 7 天内的会议室,因此导航栏显示了最近 7 天的日期。代码第 17~54 行采用了 ColorUI 中的 card 样式,包括申请人信息(姓名和单位短号)、会议时间、会议地点、会议主题和预约单提交时间等信息。代码第 48~51 行显示分割线。代码第 56~58 行对应显示"会议室预约"按钮。用户现在可以在该页面中查看该会议室最近 7 天内的预约信息,从而选择预约哪天的会议室。

这里需要在 room.wxss 引入如下样式:

```
1   page{
2     background:white;
3     padding-bottom:110rpx;
4   }
5   scroll-view{
6     height:128rpx;
7
8     width:100%;
9   }
10  scroll-view .list{
11    display:flex;
12    flex-wrap:nowrap;
13    justify-content:flex-start;
14      width:1302rpx;
15  }
16  scroll-view .listItem{
17    text-align:center;
18    width:186rpx;
19    height:128rpx;
20    background-color:rgba(255,0,0,0.137);
21    padding-top:30rpx;
22    box-sizing:border-box;
```

```css
23      /* float: left; */
24      display: inline-block;
25  }
26  scroll-view .listItem text{
27      display: block;
28  }
29  scroll-view .listItem .name{
30      font-size: 30rpx;
31  }
32  scroll-view .listItem .date{
33      font-size: 30rpx;
34  }
35  scroll-view .current{
36      background-color: rgba(255, 0, 0, 0.473);
37  }
38  scroll-view .current text{
39      color: #fff;
40  }
41  .operate{
42      display: flex;
43      flex-wrap: nowrap;
44      justify-content: space-around;
45      width: 100%;
46      padding: 5rpx 0;
47      background-color:white;
48      position: fixed;
49      bottom: 0;
51  }
52  .divLine{
53  background: #E0E3DA;
54  width: 100%;
55  height: 5rpx;
56  }
```

会议室预约情况页面效果如图12-5所示。相应的room.js代码如下：

```javascript
1   var util = require('../../utils/util.js')
2   const db = wx.cloud.database()
3   Page({
4     data: {
5       calendar: [],
6       width: 0,
7       currentIndex: 0,
8       roomID: "",
9       selectDate: "",
10      records:[]
11    },
12
13    onLoad: function(options) {
```

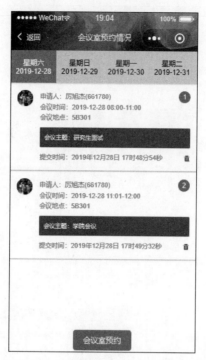

图 12-5 会议室预约情况页面效果

```
14      var roomID = options.roomID
15      var date1 = new Date()
16      var weeks_ch = ['日','一','二','三','四','五','六'];
17      var calendar = []
18      for (var i = 0; i < 7; i++) {
19        var myDate = {}
20        var date2 = new Date(date1);
21        date2.setDate(date1.getDate() + i);
22        myDate.date = util.formatTime(date2)
23        myDate.week = '星期' + weeks_ch[date2.getDay()]
24        calendar.push(myDate)
25      }
26      this.setData({
27        width: 185 * 7, calendar: calendar, roomID: roomID, selectDate: calendar[0].date
28      })
29    },
30
31    getRecords()
32    {
33      var roomID = this.data.roomID
34      var selectDate = this.data.selectDate
35      db.collection('order').where({
36        roomID: roomID,
37        selectDate: selectDate
38      }).get().then(res =>{
```

```javascript
      this.setData({ records: res.data })
    })
  },

  onShow:function(event)
  {
    this.getRecords()
  },

  select: function(event) {
    this.setData({
      currentIndex: event.currentTarget.dataset.index,
      selectDate: event.currentTarget.dataset.date
    })
    this.getRecords()
  },

  gotoAppointment: function(event) {
    var phone = wx.getStorageSync('phone');
    if (phone == "") {
      wx.navigateTo({
        url: '../login/login'
      })
    } else {
      db.collection('contact').where({
        celphone: parseInt(phone)
      }).get().then(res =>{
        console.log(res)
        if(res.data.length == 1)
        {
          wx.setStorageSync('user', res.data[0]);
          var roomID = this.data.roomID
          var selectDate = this.data.selectDate
          wx.navigateTo({
            url: '../meetingform/meetingform? roomID = ' + roomID + '&selectDate = ' + selectDate
          })
        }
        else
        {
          wx.showToast({
            title: '您没有权限预约',
            icon: 'success',
            duration: 2000
          })
        }
      })
    }
  },
```

```
89      deleteRecord:function(event)
90      {
91        var id = event.currentTarget.dataset.index
92        var records = this.data.records
93        var phone = wx.getStorageSync('phone');
94        if (phone == records[id].adminphone || phone == records[id].celphone) {
95          wx.cloud.callFunction({
96            name: 'DeleteOrder',
97            data: {
98              id: records[id]._id,
99            }
100         }).then(res =>{
101           records.splice(id, 1)
102           this.setData({
103             records: records
104           })
105           wx.showToast({
106             title: '删除成功',
107             icon: 'success',
108             duration: 2000
109           })
110         }).catch(err =>{
111         })
112       }
113       else
114       {
115         wx.showToast({
116           title: '您没有权限删除该预约',
117           icon: 'success',
118           duration: 2000
119         })
120       }
121     },
122   })
```

代码第 13~29 行在页面加载页面中计算最近 7 天的日期信息。其中，代码第 15 行对应获取今日日期 date1，代码第 18~25 行以今日日期 date1 为准，计算 7 天内的日期信息加入 calendar 数组。代码第 31~41 行获取会议室预约单中 roomID 字段为该会议室房间号，预约日期为用户选择的预约日期的预约单。代码第 43~46 行在 onShow 事件中获取会议室预约信息，由于用户填写会议室预约单后使用 wx.navigateBack 回到此页面不会调用 onLoad 事件，因此获取会议室预约信息放在 onShow 事件中，代码第 48~55 行对应用户选择日期导航栏事件，用户选择相应的日期，代码第 54 行重新获取该日期对应的会议室预约单。代码第 57~87 行对应单击"会议室预约"按钮事件，其中代码第 58~63 行判断用户是否登录，如果没有登录则跳转到登录页面，代码第 64~76 行在数据库集合 contact 中读取手机号为 phone 的用户信息，如果不存在此用户，表示该用户没有权限预约会议室，如果存在此用户，则把用户信息存入缓存中。代码第 89~121 行对应每条记录中的删除图标事件，单击该图标可以删除该预约单,这里限定只有本人或者会议室管理员才有权限对预约单进行删除,代码第 94 行确认登录用户是否是该预约单的申请者或者是会议室管理员,由于管理员也有权限删除该预约单,因此删除操作需要放在云函数中,删除预约单后,需要把该预约单从数组 records 中删除。

相应的云函数 DeleteOrder 代码如下:

```
1   const cloud = require('wx-server-sdk')
2   cloud.init()
3   const db = cloud.database()
4   exports.main = async (event, context) =>{
5     var id = event.id
6     try {
7       return await db.collection('order').doc(id).remove()
8     } catch (e) {
9       console.error(e)
10    }
11  }
```

云函数 DeleteOrder 执行从数据库 order 中删除_id 的记录。

12.3　填写预约单页面

填写预约单页面采用 ColorUI 中的表单样式,用微信开发者工具打开 ColorUI 中的 demo 文件,在 tabBar 中选择"组件"选项,然后单击"表单",进入"表单"页面,选择的文件上传组件样式(代码见 pages/component/form),选择的表单样式如图 3-36 所示,填写预约单页面 meetingform.wxml 代码如下:

```
1   <cu-custom bgImage="/images/sylb.jpg" isBack="{{true}}">
2     <view slot="backText">返回</view>
3     <view slot="content">填写预约单</view>
4   </cu-custom>
5
6   <image style="width:100%" mode='scaleToFill' src="{{order.url}}" bindtap="showImg"></image>
7   <form>
8     <view class="cu-bar bg-white solid-bottom">
9       <view class="action">
10        <text class="cuIcon-title text-orange"></text>会议室和申请人信息:
11      </view>
12    </view>
13    <view class="cu-form-group">
14      <view class="title">
15        会议室:</view>
16      <input value='{{order.roomID}}' disabled></input>
17    </view>
18    <view class="cu-form-group">
19      <view class="title">
```

```
20      申请日期:</view>
21      <input value = '{{order.selectDate}}' disabled></input>
22    </view>
23    <view class = "cu-form-group">
24      <view class = "title">
25      容纳人数:</view>
26      <input value = '{{order.capacity}}人' disabled></input>
27    </view>
28    <view class = "cu-form-group">
29      <view class = "title">
30      管理员:</view>
31      <input value = '{{order.adminname}}' disabled></input>
32    </view>
33    <view class = "cu-form-group">
34      <view class = "title">
35      管理员短号:</view>
36      <input value = '{{order.adminphone_cornet}}' disabled></input>
37    </view>
38    <view class = "cu-form-group">
39      <view class = "title">
40      申请人姓名:</view>
41      <input value = '{{order.name}}' disabled></input>
42    </view>
43    <view class = "cu-form-group">
44      <view class = "title">
45      申请人短号:</view>
46      <input value = '{{order.phone_cornet}}' disabled></input>
47    </view>
48
49    <view class = "cu-bar bg-white solid-bottom margin-top">
50      <view class = "action">
51        <text class = "cuIcon-title text-orange"></text>申请者填写:
52      </view>
53    </view>
54    <view class = "cu-form-group">
55      <view class = "title">
56        <text style = "color:red">*</text>会议主题:</view>
57      <input bindinput = 'updateValue' data-name = 'title' value = '{{order.title}}' placeholder =
58  "请输入会议主题"></input>
59    </view>
60    <view class = "cu-form-group">
61      <view class = "title">
62        <text style = "color:red">*</text>会议开始时间:</view>
63      <picker mode = "time" data-name = 'startTime' value = "{{order.startTime}}" start = "08:00" end = "22:00"
64  bindchange = "TimeChange">
65        <view class = "picker">
66          {{order.startTime}}
67        </view>
68      </picker>
```

```
69      </view>
70      <view class = "cu-form-group">
71        <view class = "title">
72          <text style = "color:red"> * </text>会议结束时间:</view>
73        <picker mode = "time" data-name = 'endTime' value = "{{order.endTime}}" start = "08:00" end = "22:00"
74    bindchange = "TimeChange">
75          <view class = "picker">
76            {{order.endTime}}
77          </view>
78        </picker>
79      </view>
80    </form>
81
82    <view class = "flex padding justify-center">
83      <button class = "cu-btn lg bg-green" bindtap = "submitform">提交预约</button>
84    </view>
```

代码第 6 行显示会议室照片,用户可以查看该照片对会议室的情况有所了解。表单分 2 部分:第 1 部分对应代码第 8～47 行,显示会议室和申请人信息,这部分不需要用户填写,包括会议室编号、申请日期、会议室容纳人数、管理员信息和申请人信息;第 2 部分对应代码第 49～80 行,显示会议主题、会议开始时间和会议结束时间,在这里尽可能让用户少填写信息,能够快速完成会议室预约申请,填写预约单页面效果如图 12-6 所示,因为页面较长,会议室图片和"提交预约"按钮并未显示出来。

图 12-6 填写预约单页面效果

填写完预约单后，会在数据库 order 集合中插入预约记录，数据库 order 集合中的字段信息如图 12-7 所示。

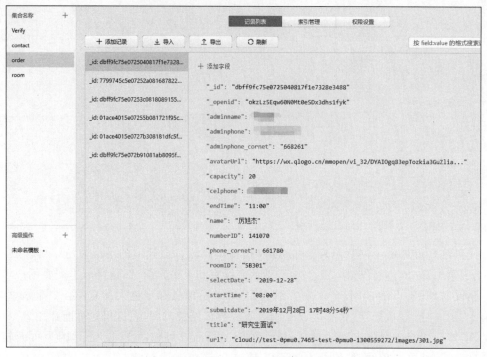

图 12-7　数据库 order 集合中的字段信息

相应的 meetingform.js 代码如下：

```
const db = wx.cloud.database()
Page({
  data: {
    order: {
      startTime: '08:00',
      endTime: '11:00',
      title: ''
    },
  },

  onLoad: function(options) {
    var roomID = options.roomID
    var selectDate = options.selectDate
    var order = this.data.order
    order.roomID = roomID
    order.selectDate = selectDate
    var user = wx.getStorageSync('user');
    order.name = user.name
    order.phone_cornet = user.phone_cornet
    order.celphone = user.celphone
    order.numberID = user.student_number
```

```js
22      var userInfo = wx.getStorageSync('userInfo');
23      order.avatarUrl = userInfo.avatarUrl
24      db.collection('room').where({
25        roomID: roomID
26      }).get().then(res =>{
27        var result = res.data[0]
28        order.capacity = result.capacity
29        order.adminname = result.adminname
30        order.adminphone = result.adminphone
31        order.adminphone_cornet = result.adminphone_cornet
32        order.url = result.url
33        this.setData({
34          order: order
35        })
36      })
37    },
38
39    updateValue: function(event) {
40      let name = event.currentTarget.dataset.name;
41      let order = this.data.order
42      order[name] = event.detail.value
43      this.setData({
44        order: order
45      })
46    },
47
48    TimeChange(event) {
49      let name = event.currentTarget.dataset.name;
50      var order = this.data.order
51      order[name] = event.detail.value
52      this.setData({
53        order: order
54      })
55    },
56
57    submitform: function(event) {
58      var order = this.data.order
59      var order = this.data.order
60      var roomID = order.roomID
61      var selectDate = order.selectDate
62      var status = 1
63      if (order.title == '') {
64        wx.showToast({
65          title: "主题不能为空!",
66          icon: 'success',
67          duration: 2000
68        })
69      } else {
70        if (order.startTime >= order.endTime) {
```

```
71          wx.showToast({
72            title: "结束时间不对!",
73            icon: 'success',
74            duration: 2000
75          })
76        } else {
77          db.collection('order').where({
78            roomID: roomID,
79            selectDate: selectDate
80          }).get().then(res =>{
81            var data = res.data
82            for (var i = 0; i < data.length; i++) {
83              if (data[i].endTime < order.startTime || order.endTime < data[i].startTime) {
84                console.log("没有冲突");
85              } else {
86                status = 0
87              }
88            }
89            if (status == 1) {
90              const date = new Date();
91              const year = date.getFullYear()
92              const month = date.getMonth() + 1
93              const day = date.getDate()
94              const hour = date.getHours()
95              const minute = date.getMinutes()
96              const second = date.getSeconds()
97              var datastring = year + '年' + month + '月' + day + '日 ' + hour + '时' +
98  minute + '分' + second + '秒';
99              order.submitdate = datastring
100             db.collection('order').add({
101               data: order
102             })
103               .then(res =>{
104                 console.log(res)
105                 wx.navigateBack({
106                   delta: 1
107                 })
108               })
109           }
110           else
111           {
112             wx.showToast({
113               title: "会议时间冲突!",
114               icon: 'success',
115               duration: 2000
116             })
117           }
118         })
119       }
```

```
120      }
121    }
122  })
```

代码第 11～37 行对应页面加载事件,其中代码第 17～21 行读取缓存中 user(用户信息)数据,代码第 22、23 行读取 userInfo(用户微信信息)数据,代码第 24～26 行根据 roomID 查询会议室信息。代码第 39～46 行对应用户输入会议主题。代码第 48～55 行对应用户选择会议开始时间和会议结束时间。代码第 57～122 行对应单击"提交预约"按钮事件,其中代码第 63～68 行判断用户是否输入会议主题,代码第 70～75 行对应检测会议结束时间是否大于会议开始时间,代码第 77～109 行对应检测用户填写的预约单开始时间和结束时间是否和已有的预约单时间冲突,代码第 77～80 行读取数据库 order 集合中房间号为 roomID、日期为 selectDate 的记录,代码第 82～88 行检测是否存在时间冲突。冲突检测的逻辑思路如下:a、b 代表一个时间段,x、y 代表一个时间段(a、b、x、y 都是时间戳,所以 a<b,x<y)。2 个时间段不冲突有两种情况:情况 1,y<a;情况 2,b<x。代码第 83 行采用这种思路检测是否存在时间冲突,如果没有冲突,则把该预约单添加到 order 集合中,代码第 100～102 行对应把记录添加到数据库中。

12.4 我的预约页面

我的预约页面的样式和会议室预约情况中的 card 样式一样,我的预约页面 myorder.wxml 代码如下:

```
1   <cu-custom bgImage = "/images/sylb.jpg" isBack = "{{true}}">
2     <view slot = "backText">返回</view>
3     <view slot = "content">我的预约</view>
4   </cu-custom>
5
6   <view class = "cu-list menu-avatar comment solids-top">
7     <block wx:for = "{{records}}" wx:key = "index">
8       <view class = "cu-item">
9         <view class = "cu-avatar round" style = "background-image:url({{item.avatarUrl}});"></view>
10        <view class = "content">
11          <view class = "flex justify-between">
12            <view class = "text-gray text-df">申请人:{{item.name}}({{item.phone_cornet}})</view>
13            <view>
14              <view class = "cu-avatar round sm bg-green">{{index + 1}}</view>
15            </view>
16          </view>
17          <view class = "text-gray text-content text-df">
18            会议时间:{{item.selectDate}} {{item.startTime}}-{{item.endTime}}
19          </view>
20          <view class = "text-gray text-content text-df">
21            会议地点:{{item.roomID}}
22          </view>
```

```
23          <view class="bg-pink padding-sm radius margin-top-sm text-sm">
24            <view class="flex">
25              <view class="flex-sub">会议主题：{{item.title}}</view>
26            </view>
27          </view>
28          <view class="margin-top-sm flex justify-between">
29            <view class="text-gray text-df">提交时间：{{item.submitdate}}</view>
30            <view>
31              <text class="cuIcon-deletefill text-red margin-left-sm" data-index="{{index}}"
32  bindtap="deleteRecord"></text>
33            </view>
34          </view>
35        </view>
36      </view>
37      <view class="container">
38        <!-- 分割线 -->
39        <view class="divLine"></view>
40      </view>
41    </block>
42  </view>
```

每个预约单记录样式和 room.wxml 中的样式一样，这里不再重复说明，我的预约页面效果如图 10-8 所示。

图 12-8　我的预约页面效果

相应的 myorder.js 代码如下：

```
1   const db = wx.cloud.database()
2   Page({
3     data: {
4       records: [],
5       page: 0,
6     },
7   
8     onLoad: function (options) {
9       var phone = wx.getStorageSync('phone');
10      console.log(phone)
11      db.collection('order').where({
12        celphone: parseInt(phone)
13      }).orderBy('selectDate', 'desc').get().then(res =>{
14        this.setData({ records: res.data })
15      })
16    },
17  
18    deleteRecord: function (event) {
19      var id = event.currentTarget.dataset.index
20      var records = this.data.records
21      var phone = wx.getStorageSync('phone');
22      if (phone == records[id].adminphone || phone == records[id].celphone) {
23        wx.cloud.callFunction({
24          name: 'DeleteOrder',
25          data: {
26            id: records[id]._id,
27          }
28        }).then(res =>{
29          records.splice(id, 1)
30          this.setData({
31            records: records
32          })
33          wx.showToast({
34            title: '删除成功',
35            icon: 'success',
36            duration: 2000
37          })
38        }).catch(err =>{
39          // handle error
40        })
41      }
42      else {
43        wx.showToast({
44          title: '您没有权限删除该预约',
45          icon: 'success',
46          duration: 2000
47        })
```

```
48        }
49      },
50
51      onReachBottom: function () {
52        var page = this.data.page + 20
53        var phone = wx.getStorageSync('phone');
54        db.collection('order').where({
55          celphone: parseInt(phone)
56        }).orderBy('selectDate', 'desc').skip(page).get().then(res =>{
57          var records = this.data.records
58          records = records.concat(res.data)
59          this.setData({ records: records, page: page })
60        })
61      },
62    })
```

代码第 8～16 行对应页面加载事件,从数据库 order 集合中读取 celphone 字段值为 phone 的所有记录,因为微信小程序端一次最多只能读取 20 条记录,所以需要配合触底刷新 onReachBottom 事件,读取超过 20 条记录的数据。代码第 18～49 行对应删除预约单,该功能和 room.js 中的功能相同。代码第 51～61 行对应触底刷新 onReachBottom 事件,结合 Collection.skip 指令进行多次读取。

第 13 章

AI+微信小程序

如今 AI 浪潮席卷全球，AI 可以帮助计算机以一种自然的、更接近人类的方式理解这个世界，其中一个令人兴奋的新进展就是计算机视觉技术，它能让计算机直接理解输入的视觉信息，并由此"看到"这个世界。

百度 AI 开放平台集结了图像技术、人脸与人体识别、语音技术、自然语言处理等 AI 技术。本章以百度 AI 为例，演示微信小程序使用百度 AI 实现车牌识别和通用物体识别。

AI＋微信小程序包含 3 个页面：

```
1  "pages":[
2    "pages/home/home",                              //项目主页
3    "pages/carAi/carAi",                            //车牌识别页面
4    "pages/objectRecognition/objectRecognition"     //通用物体识别页面
5  ],
```

13.1 百度 AI 开放平台

1. 成为开发者

进入百度 AI 开放平台（https://ai.baidu.com/），单击导航右侧的控制台进行用户登录，如果未注册则先进行用户注册，首次使用时，登录后将会进入开发者认证页面，填写相关信息完成开发者认证（如果之前已经是百度云用户或百度开发者中心用户，此步可略过）。通过控制台左侧导航，选择"产品服务"→"人工智能"选项，进入具体 AI 开放平台服务项控制面板（如文字识别、人脸识别），进行相关业务操作，如图 13-1 所示。

2. 创建应用

账号登录成功后，需要创建应用才可正式调用 AI 能力。应用是用户调用 API 服务的基本操作单元，可以基于应用创建成功后获取的 API Key 及 Secret Key 进行接口调用操作

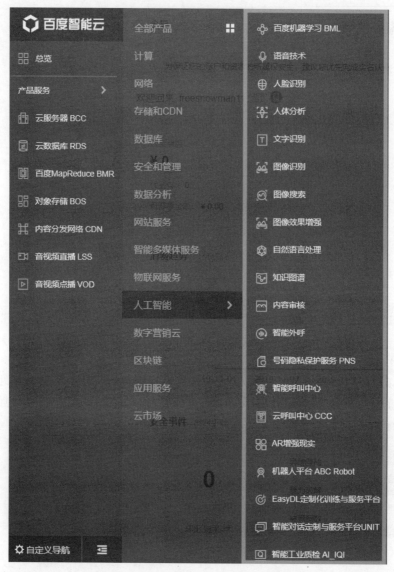

图 13-1 百度 AI 开放平台服务项控制面板

及相关配置。本章使用了百度图像识别功能,如图 13-1 所示,单击"图像识别"选项,开始创建应用,如图 13-2 所示。

在创建新应用时,用户需要勾选"图像识别"复选框(默认勾选)、"文字识别"复选框(车牌识别),如图 13-3 所示。

3. 获取密钥

在创建完应用后,平台将会分配此应用的相关凭证,如图 13-4 所示,主要为 AppID、API Key、Secret Key。以上 3 个信息是应用实际开发的主要凭证,每个应用之间各不相同,请妥善保管。

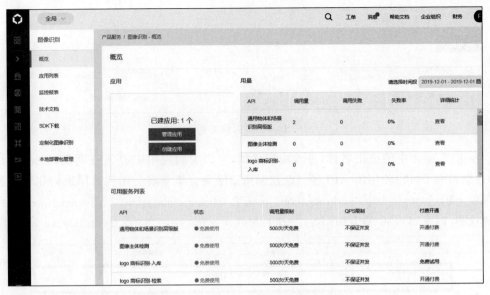

图 13-2 在百度 AI 开放平台创建应用

图 13-3 创建新应用

	应用名称	AppID	API Key	Secret Key	创建时间	操作
1	微信小程序演示	17900524	Y3yP9kPsiA6f7383p9G6bBGh	****** 显示	2019-12-01 20:37:47	报表 管理 删除

图 13-4　应用相关凭证

单击图 13-4 中的应用名称，进入应用详情，可以看到应用中的 API 列表，如图 13-5 所示，这里列举了可以使用的 API 接口以及调用量限制。本章将会用到通用物体和场景识别高级版 API(https://aip.baidubce.com/rest/2.0/image-classify/v2/advanced_general)和车辆检测(即车牌识别 API(https://aip.baidubce.com/rest/2.0/ocr/v1/license_plate)。

API	状态	请求地址	调用量限制	QPS限制
通用物体和场景识别高级版	●免费使用	https://aip.baidubce.com/rest/2.0/image-classify/v2/advanced_general	500次/天免费	不保证并发
图像主体检测	●免费使用	https://aip.baidubce.com/rest/2.0/image-classify/v1/object_detect	500次/天免费	不保证并发
logo 商标识别-入库	●免费使用	https://aip.baidubce.com/rest/2.0/realtime_search/v1/logo/add	500次/天免费	不保证并发
logo 商标识别-检索	●免费使用	https://aip.baidubce.com/rest/2.0/image-classify/v2/logo	500次/天免费	不保证并发
logo 商标识别-删除	●免费使用	https://aip.baidubce.com/rest/2.0/realtime_search/v1/logo/delete	500次/天免费	不保证并发
菜品识别	●免费使用	https://aip.baidubce.com/rest/2.0/image-classify/v2/dish	500次/天免费	不保证并发
车型识别	●免费使用	https://aip.baidubce.com/rest/2.0/image-classify/v1/car	500次/天免费	不保证并发
动物识别	●免费使用	https://aip.baidubce.com/rest/2.0/image-classify/v1/animal	500次/天免费	不保证并发
植物识别	●免费使用	https://aip.baidubce.com/rest/2.0/image-classify/v1/plant	500次/天免费	不保证并发
果蔬识别	●免费使用	https://aip.baidubce.com/rest/2.0/image-classify/v1/classify/ingredient	500次/天免费	不保证并发
自定义菜品识别-入库	●免费使用	https://aip.baidubce.com/rest/2.0/image-classify/v1/realtime_search/dish/add	无限制	不保证并发
自定义菜品识别-删除	●免费使用	https://aip.baidubce.com/rest/2.0/image-classify/v1/realtime_search/dish/delete	无限制	不保证并发
自定义菜品识别-检索	●免费使用	https://aip.baidubce.com/rest/2.0/image-classify/v1/realtime_search/dish/search	500次/天免费	不保证并发
地标识别	●免费使用	https://aip.baidubce.com/rest/2.0/image-classify/v1/landmark	500次/天免费	不保证并发
车辆检测	●免费使用	https://aip.baidubce.com/rest/2.0/image-classify/v1/vehicle_detect	500次/天免费	不保证并发

图 13-5　应用中的 API 列表

4. 鉴权认证机制

百度 AI 开放平台使用 OAuth2.0 授权调用开放 API，调用 API 时必须在 URL 中带上 access_token 参数。获取 Access Token 的流程如下：

向授权服务地址 https://aip.baidubce.com/oauth/2.0/token 发送请求（推荐使用 POST），并在 URL 中带上以下参数。

- grant_type：必需参数，固定为 client_credentials；
- client_id：必需参数，应用的 API Key；
- client_secret：必需参数，应用的 Secret Key；

例如：

> https://aip.baidubce.com/oauth/2.0/token? grant _ type = client _ credentials&client _ id = Y3yP9kPsiA6f7383p9G6bBGh&client_secret = LO29q8OwKEuwvG2OIyAGTmQKUM4PTyx5&

读者如果想要测试是否能够获取 Access Token，可以把上面的 HTTPS 请求中的 client_id 替换为自己的 API Key，client_secret 替换为自己的 Secret Key，替换后把上面的 HTTPS 请求复制到浏览器地址栏中，按 Enter 键就可以看到返回的结果。通过浏览器获取 Access Token 如图 13-6 所示。

图 13-6　通过浏览器获取 Access Token

服务器返回的 JSON 文本参数如下：

- access_token：要获取的 Access Token；
- expires_in：Access Token 的有效期（单位为秒，一般为 1 个月）；
- 其他参数忽略，暂时不用。

从图 13-6 中可以看到成功获取了 Access Token，expires_in 的值为 259200，表示有效期为 30 天。

13.2 百度 AI 开放平台接口测试

本节使用 Postman 来测试百度 AI 开放平台接口。

1. 获取 Access Token

在 Postman 中输入地址 https://aip.baidubce.com/oauth/2.0/token,选择 POST 方法,在 Params 中输入 grant_type、client_id、client_secret 这 3 个参数,单击 Send 按钮发送 HTTP 请求,在下面的 Body 中可以看到百度 AI 开放平台返回的 Access Token 值,使用 Postman 获取 Access Token 的结果如图 13-7 所示。

图 13-7 使用 Postman 获取 Access Token 的结果

2. 文字识别——调用车牌识别接口

1) 接口描述

对机动车蓝牌、绿牌、单/双行黄牌的地域编号和车牌号进行识别,并能同时识别图像中的多张车牌。

2) 请求说明

HTTP 方法：POST。

请求 URL：https://aip.baidubce.com/rest/2.0/ocr/v1/license_plate。

URL 参数(见表 13-1)：

表 13-1 URL 参数 1

参　　数	值
access_token	通过 API Key 和 Secret Key 获取的 access_token

Header 参数(见表 13-2)：

表 13-2　Header 参数 1

参　　数	值
Content-Type	application/x-www-form-urlencoded

Body 中请求参数（见表 13-3）：

表 13-3　Body 中请求参数 1

参　　数	类　　型	是否必须	说　　明
image	string	是	图像数据，Base64 编码后进行 urlencode，要求 Base64 编码和 urlencode 后大小不超过 4MB，最短边至少为 15px，最长边最大为 4096px，支持 JPG/JPEG/PNG/BMP 格式
multi_detect	boolen	否	是否检测多张车牌，默认为 false，当置为 true 时可以对一张图片内的多张车牌进行识别

车牌识别接口必须要提供一个 image 参数，该参数是 Base64 的图像数据，这里使用在线图片转换 Base64 方法获取 image 数据（http://imgbase64.duoshitong.com/），如图 13-8 所示，上传图片获得 Base64 数据，这里需要提取"data:image/jpeg;base64,"之后的数据。

图 13-8　在线图片转换 Base64 数据

在 Postman 工具中新建 HTTP 请求，Postman 测试车辆识别接口如图 13-9 所示，采用 POST 方式，输入网址 https://aip.baidubce.com/rest/2.0/ocr/v1/license_plate，在 Params 中输入 access_token 的值（获取方式见图 13-7），在 Body 项中输入 image 参数，值为图片 Base64 数据（获取方式见图 13-8），单击 Send 按钮发送 HTTP 请求，获得车辆识别的结果。

3. 图像识别——调用通用物体及场景接口

1）接口描述

该请求用于通用物体及场景识别，即对于输入的一张图片（可正常解码，且长宽比适宜），输出图片中的多个物体及场景标签。

2）请求说明

HTTP 方法：POST。

请求 URL：https://aip.baidubce.com/rest/2.0/image-classify/v2/advanced_general。

URL 参数（见表 13-4）：

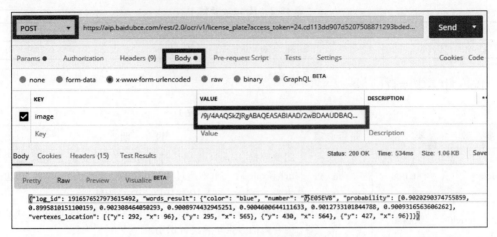

图 13-9　Postman 测试车辆识别接口

表 13-4　URL 参数 2

参　　数	值
access_token	通过 API Key 和 Secret Key 获取的 access_token

Header 参数（见表 13-5）：

表 13-5　Header 参数 2

参　　数	值
Content-Type	application/x-www-form-urlencoded

Body 中请求参数（见表 13-6）：

表 13-6　Body 中请求参数 2

参　　数	是否必选	类　　型	可选值范围	说　　明
image	是	string	—	Base64 编码字符串，以图片文件形式请求时必填（支持图片格式：JPG、BMP、PNG、JPEG），图片大小不超过 4MB。最短边至少为 15px，最长边最大为 4096px。注意，图片需要 Base64 编码、去掉编码头后再进行 urlencode
baike_num	否	integer	0~5	返回百科信息的结果数，默认为 0，不返回；2 为返回前 2 个结果的百科信息，以此类推

通用物体及场景接口必须要提供一个 image 参数，该参数是 Base64 的图像数据，图片转换 Base64 方法如图 13-8 所示。

在 Postman 工具中新建 HTTP 请求，Postman 测试通用物体及场景接口如图 13-10 所示，采用 POST 方式，输入网址 https://aip.baidubce.com/rest/2.0/image-classify/v2/advanced_general，在 Params 中输入 access_token 的值（获取方式见图 13-7），在 Body 项中输入 image 参数，值为图片 Base64 数据（获取方式见图 13-8），单击 Send 按钮发送 HTTP 请求，获得通用物体及场景的结果。

图 13-10　Postman 测试通用物体及场景接口

13.3　项目主页

项目主页提供了两个功能：车牌识别（文字识别）和物体识别（图像识别）。

项目主页 home.wxml 的代码如下：

```
1   <cu-custom bgColor = "bg-gradual-pink" isBack = "{{false}}">
2     <view slot = "content">主页面</view>
3   </cu-custom>
4
5   <view class = "cu-bar bg-white solid-bottom">
6     <view class = "action">
7       <text class = "cuIcon-title text-orange"></text>应用列表
8     </view>
9   </view>
10  <view class = "cu-list grid col-3">
11    <view class = "cu-item" wx:for = "{{boxlist}}" wx:key = "index" wx:for-item = "item">
```

```
12      <view data-id='{{item.name}}' bindtap='enterapplication'>
13        <image style="width: 40px; height: 40px" src='{{item.url}}'></image>
14        <text>{{item.name}}</text>
15      </view>
16    </view>
17  </view>
```

项目主页效果如图 13-11 所示,包含车牌识别(文字识别)和物体识别(图像识别)功能。相应的 home.js 代码如下:

图 13-11　AI+微信小程序项目主页效果

```
1   const db = wx.cloud.database()
2   Page({
3     data: {
4       boxlist:
5         [
6           {
7             name: "车牌识别",
8             url: "/images/chepai.png"
9           },
10          {
11            name: "物体识别",
12            url: "/images/wupin.png"
13          }
14        ],
15    },
```

```
16
17      enterapplication: function (event) {
18        var id = event.currentTarget.dataset.id
19        switch (id) {
20          case '车牌识别':
21            wx.navigateTo({ url: '../carAi/carAi' })
22            break;
23          case '物体识别':
24            wx.navigateTo({ url: '../objectRecognition/objectRecognition' })
25            break;
26          default:
27            break;
28        }
29      },
30    })
```

为了直观显示,本页面只有两个功能,代码第 4～15 行将这两项功能数据直接写在了 data 中,实际开发中应该写入云数据库,图片也应该上传到云存储中。代码第 17～29 行用 switch 语句判断用户单击了哪个功能。

13.4 车牌识别页面

车牌识别页面 carAi.wxml 代码如下:

```
1    <cu-custom bgColor = "bg-gradual-pink" isBack = "{{true}}">
2      <view slot = "backText">返回</view>
3      <view slot = "content">车辆识别</view>
4    </cu-custom>
5    
6    <view style = "width: 100%; height:calc(100vh - 350rpx);">
7      <camera style = "width: 100%; height:100%;" wx:if = "{{isCamera}}" device-position = 
         "back" flash = "off" binderror = "error"></camera>
8    
9      <image style = "width: 100%; height:100%;" wx:else mode = "widthFix" src = "{{src}}"></image>
10   </view>
11   <view class = "flex padding justify-center">
12     <view class = "cu-avatar round xl bg-green" bindtap = "takePhoto">
13       <view class = "text-white text-lg">{{btnTxt}}</view>
14     </view>
15   </view>
16   
17   <view class = "cu-modal {{isShow?'show':''}}" bindtap = "hideModal">
18     <view class = "cu-dialog" catchtap>
19       <radio-group class = "block" bindchange = "radioChange">
20         <view class = "cu-list menu text-left">
21           <view class = "cu-item">
22             <label class = "flex justify-between align-center flex-sub">
23               <view class = "flex-sub">颜色:{{results.color}}</view>
```

```
24              </label>
25              <label class="flex justify-between align-center flex-sub">
26                <view class="flex-sub">车牌:{{results.number}}</view>
27              </label>
28            </view>
29          </view>
30        </radio-group>
31      </view>
32    </view>
```

代码第 6～10 行使用系统相机组件。代码第 11～15 行显示"拍照"按钮。代码第 17～31 行弹出模态框显示车牌识别结果，包括颜色和车牌。车牌识别页面效果如图 13-12 所示，用户单击"拍照"按钮，识别出车牌信息。

图 13-12　车牌识别页面效果

相应的 carAi.js 代码如下：

```js
Page({
  data: {
    accessToken: "",
    isShow: false,
    results: [],
    src: "",
    isCamera: true,
    btnTxt: "拍照"
  },

  accessTokenFunc: function() {
    var that = this
    wx.cloud.callFunction({
      name: 'baiduAccessToken',
      success: res =>{
        console.log(res)
        that.data.accessToken = res.result.access_token
        wx.setStorageSync("access_token", res.result.access_token)
        wx.setStorageSync("time", new Date().getTime())
      },
      fail: err =>{
        wx.showToast({
          icon: 'none',
          title: '调用失败',
        })
        console.error(err)
      }
    })
  },

  onLoad() {
    this.ctx = wx.createCameraContext()
    var time = wx.getStorageSync("time")
    var curTime = new Date().getTime()
    var timeNum = new Date(parseInt(curTime - time)).getDate()
    console.log("=======" + timeNum)
    var accessToken = wx.getStorageSync("access_token")
    console.log("====accessToken===" + accessToken + "a")
    if (timeNum > 28 || (accessToken == "" ||
        accessToken == null || accessToken == undefined)) {
      this.accessTokenFunc()
    } else {
      this.setData({
        accessToken: wx.getStorageSync("access_token")
      })
    }
  },

```

```js
49      takePhoto() {
50        var that = this
51        if (this.data.isCamera == false) {
52          this.setData({
53            isCamera: true,
54            btnTxt: "拍照"
55          })
56          return
57        }
58        this.ctx.takePhoto({
59          quality: 'normal',
60          success: (res) =>{
61            this.setData({
62              src: res.tempImagePath,
63              isCamera: false,
64              btnTxt: "重拍"
65            })
66            wx.showLoading({
67              title: '正在加载中',
68            })
69            wx.getFileSystemManager().readFile({
70              filePath: res.tempImagePath,
71              encoding: "base64",
72              success: res =>{
73                console.log("图片读取成功")
74                that.req(that.data.accessToken, res.data)
75              },
76              fail: res =>{
77                wx.hideLoading()
78                wx.showToast({
79                  title: '拍照失败,未获取相机权限或其他原因',
80                  icon: "none"
81                })
82              }
83            })
84          }
85        })
86      },
87
88      req: function(token, image) {
89        var that = this
90        wx.cloud.callFunction({
91          name: 'license_plate',
92          data: {
93            "image": image,
94            "token": token
95          },
96          success: res =>{
97            wx.hideLoading()
```

```
 98            let data = res.result
 99            console.log(JSON.stringify(res))
100            if ("words_result" in data) {
101              var results = res.result.words_result
102              if (results != undefined && results != null) {
103                that.setData({
104                  isShow: true,
105                  results: results
106                })
107                console.log(results)
108              }
109            }
110            if ("error_code" in data) {
111              wx.showToast({
112                icon: 'none',
113                title: 'AI识别失败,请重新拍照!',
114              })
115            }
116          },
117          fail: err =>{
118            wx.showToast({
119              icon: 'none',
120              title: '调用失败',
121            })
122            console.error(err)
123          }
124        })
125      },
126
127      radioChange: function (e) {
128        console.log(e.detail.value)
129      },
130
131      hideModal: function() {
132        this.setData({
133          isShow: false,
134        })
135      },
136    })
```

代码第 11~29 行调用云函数 baiduAccessToken 获取 Access Token 值,然后把 Access Token 写入本地缓存中。代码第 31~47 行的页面数据加载事件中,读取本地缓存中 Access Token 值,由于 Access Token 值有效期为 30 天,因此需要判断 Access Token 值是否过期,其中代码第 35 行计算当前时间和本地缓存中 time 值之间的天数差,如果天数超过 28 天(有效期为 30 天)或者本地缓存中没有 Access Token 值,代码第 41 行向百度 AI 开放平台重新获取 Access Token 值。代码第 49~86 行对应页面中的单击"拍照"按钮事件,调用相机的 CameraContext.takePhoto 接口拍摄照片,其中代码第 69~72 行读取拍摄的照片,并进行 Base64 编码,代码第 74 行调用百度 AI 识别平台的车辆识别接口对车牌进行识别。代码第 88～125 行调用云函数 license_plate 发送 HTTP 请求。

因为发送 HTTP 请求并没有使用云数据库和云存储,所以读者也可以先在 VS Code 中进行调试,VS Code 测试车牌识别接口如图 13-13 所示。

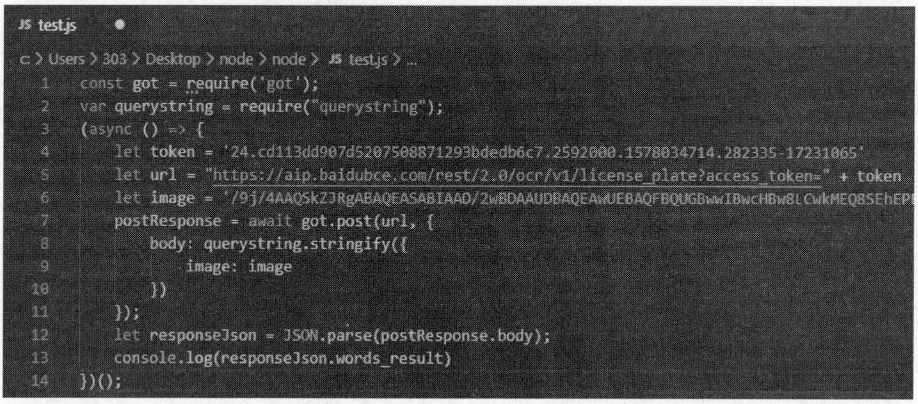

```js
const got = require('got');
var querystring = require("querystring");
(async () => {
    let token = '24.cd113dd907d5207508871293bdedb6c7.2592000.1578034714.282335-17231065'
    let url = "https://aip.baidubce.com/rest/2.0/ocr/v1/license_plate?access_token=" + token
    let image = '/9j/4AAQSkZJRgABAQEASABIAAD/2wBDAAUDBAQEAwUEBAQFBQUGBwwIBwcHBw8LCwkMEQ8SEhEPE
    postResponse = await got.post(url, {
        body: querystring.stringify({
            image: image
        })
    });
    let responseJson = JSON.parse(postResponse.body);
    console.log(responseJson.words_result)
})();
```

图 13-13　VS Code 测试车牌识别接口

云函数 baiduAccessToken 中 index.js 的代码如下:

```js
 1  const got = require('got');
 2  exports.main = async(event, context) =>{
 3    let apiKey = '8qS9tdNISiyIwsfw7jc0pHQI'
 4    let grantType = 'client_credentials'
 5    let secretKey = 'aRpQUKI3AkgDhd04BwnDHvtpg5fqeLwu'
 6    let url = 'https://aip.baidubce.com/oauth/2.0/token?client_id = ' + apiKey + '&grant_
 7  type = ' + grantType + '&client_secret = ' + secretKey
 8    let getResponse = await got(url)
 9    let responseJson = JSON.parse(getResponse.body);
10    return responseJson
11  }
```

使用 got 模块发送 HTTP GET 请求,请读者自行替换代码中的 apiKey、grantType 和 secretKey 值。

云函数 license_plate 中 index.js 的代码如下:

```js
 1  const got = require('got');
 2  const querystring = require("querystring");
 3  exports.main = async (event, context) =>{
 4    let token = event.token
 5    let url = "https://aip.baidubce.com/rest/2.0/ocr/v1/license_plate?access_token=" + token
 6    let image = event.image
 7    postResponse = await got.post(url, {
 8        body: querystring.stringify({
 9            image: image
10        })
```

```
11      });
12      let responseJson = JSON.parse(postResponse.body);
13      console.log(responseJson.words_result)
14      return responseJson
15  }
```

使用 got.post 发送 HTTP POST 请求，因为 Header 的参数 Content-Type 的值为 application/x-www-form-urlencoded，所以 Body 中的对象必须使用 querystring.stringify 进行字符串化，这一点在 got 官方网站（https://www.ctolib.com/got.html）上有说明。需要说明的是，目前云函数包过大会出现返回超时，用户需要多试几次，项目上线后不会存在这个问题。

13.5 通用物体识别页面

通用物体识别页面 objectRecognition.wxml 代码如下：

```
1   <cu-custom bgColor="bg-gradual-pink" isBack="{{true}}">
2       <view slot="backText">返回</view>
3       <view slot="content">通用物体及场景识别</view>
4   </cu-custom>
5
6   <view style="width:100%;height:calc(100vh - 350rpx);">
7   <camera style="width:100%;height:100%;" wx:if="{{isCamera}}" device-position="back"
8   flash="off" binderror="error"></camera>
9   <image style="width:100%;height:100%;" wx:else mode="widthFix" src="{{src}}"></image>
10  </view>
11  <view class="flex padding justify-center">
12      <view class="cu-avatar round xl bg-green" bindtap="takePhoto">
13          <view class="text-white text-lg">{{btnTxt}}</view>
14      </view>
15  </view>
16
17  <view class="cu-modal {{isShow?'show':''}}" bindtap="hideModal">
18      <view class="cu-dialog" catchtap>
19          <radio-group class="block" bindchange="radioChange">
20              <view class="cu-list menu text-left">
21                  <view class="cu-item" wx:for="{{results}}" wx:key="index">
22                      <label class="flex justify-between align-center flex-sub">
23                          <view class="flex-sub">{{item.keyword}}</view>
24                          <radio class="round" value="{{item.keyword}}"></radio>
25                      </label>
26                  </view>
27              </view>
28          </radio-group>
29      </view>
30  </view>
```

和车牌识别页面效果类似,代码第 6～10 行使用系统相机组件。代码第 11～15 行显示"拍照"按钮。代码第 17～30 行弹出模态框显示物体识别结果。通用物体识别页面效果如图 13-14 所示,单击"拍照"按钮,可识别物体的信息。

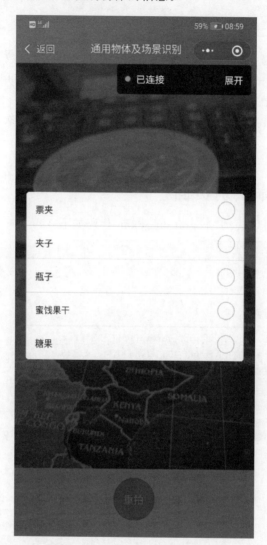

图 13-14　通用物体识别页面效果

相应的 objectRecognition.js 代码如下:

```
1  Page({
2    data: {
3      accessToken: "",
4      isShow: false,
5      results: [],
6      src: "",
7      isCamera: true,
8      btnTxt: "拍照"
```

```js
9      },
10
11     accessTokenFunc: function () {
12       var that = this
13       wx.cloud.callFunction({
14         name: 'baiduAccessToken',
15         success: res =>{
16           that.data.accessToken = res.result.access_token
17           wx.setStorageSync("access_token", res.result.access_token)
18           wx.setStorageSync("time", new Date().getTime())
19         },
20         fail: err =>{
21           wx.showToast({
22             icon: 'none',
23             title: '调用失败',
24           })
25           console.error(err)
26         }
27       })
28     },
29
30     onLoad() {
31       this.ctx = wx.createCameraContext()
32       var time = wx.getStorageSync("time")
33       var curTime = new Date().getTime()
34       var timeNum = new Date(parseInt(curTime - time)).getDate()
35       console.log("=======" + timeNum)
36       var accessToken = wx.getStorageSync("access_token")
37       console.log("==== accessToken ===" + accessToken + "a")
38       if (timeNum > 28 || (accessToken == "" ||
39         accessToken == null || accessToken == undefined)) {
40         this.accessTokenFunc()
41       } else {
42         this.setData({
43           accessToken: wx.getStorageSync("access_token")
44         })
45       }
46     },
47
48     takePhoto() {
49       var that = this
50       if (this.data.isCamera == false) {
51         this.setData({
52           isCamera: true,
53           btnTxt: "拍照"
54         })
55         return
56       }
57       this.ctx.takePhoto({
```

```
58          quality: 'normal',
59          success: (res) =>{
60            this.setData({
61              src: res.tempImagePath,
62              isCamera: false,
63              btnTxt: "重拍"
64            })
65            wx.showLoading({
66              title: '正在加载中',
67            })
68            wx.getFileSystemManager().readFile({
69              filePath: res.tempImagePath,
70              encoding: "base64",
71              success: res =>{
72                console.log("图片读取成功")
73                that.req(that.data.accessToken, res.data)
74              },
75              fail: res =>{
76                wx.hideLoading()
77                wx.showToast({
78                  title: '拍照失败,未获取相机权限或其他原因',
79                  icon: "none"
80                })
81              }
82            })
83          }
84        })
85      },
86
87      req: function (token, image) {
88        var that = this
89        wx.cloud.callFunction({
90          name: 'advanced_general',
91          data: {
92            "image": image,
93            "token": token
94          },
95          success: res =>{
96            wx.hideLoading()
97            console.log(res)
98            var num = res.result_num
99            var results = res.result.result
100           if (results != undefined && results != null) {
101             that.setData({
102               isShow: true,
103               results: results
104             })
105             console.log(results)
106           } else {
```

```
107          wx.showToast({
108            icon: 'none',
109            title: 'AI识别失败,请联系管理员',
110          })
111        }
112      },
113      fail: err =>{
114        wx.showToast({
115          icon: 'none',
116          title: '调用失败',
117        })
118        console.error(err)
119      }
120    })
121  },
122
123  radioChange: function (e) {
124    console.log(e.detail.value)
125  },
126
127  hideModal: function () {
128    this.setData({
129      isShow: false,
130    })
131  }
132 })
```

通用物体识别页面的逻辑代码和车牌识别页面的逻辑代码一样,区别在于发送的 HTTP 请求接口不同,代码第 87~121 行调用了 advanced_general 云函数发送通用物体识别请求。

云函数 advanced_general 中 index.js 的代码如下:

```
1  const got = require('got');
2  const querystring = require("querystring");
3  exports.main = async (event, context) =>{
4    let token = event.token
5    let url = "https://aip.baidubce.com/rest/2.0/image-classify/v2/advanced_general?access_token=" + token
6
7    let image = event.image
8    postResponse = await got.post(url, {
9      body: querystring.stringify({
10        image: image
11     })
12   });
13   let responseJson = JSON.parse(postResponse.body);
14   console.log(responseJson.words_result)
15   return responseJson
16 }
```

云函数 advanced_general 调用了百度 AI 开放平台通用物体识别接口 https://aip.baidubce.com/rest/2.0/image-classify/v2/advanced_general。

这里需要说明的是，车牌识别和通用物体识别都采用在云函数中发送 HTTP 请求，这样做的好处是，接口的域名不用在微信公众平台上进行配置，但是由于图片数据需要从微信小程序端发送到云函数，云函数再发送请求到百度 AI 开放平台相应接口，增加了请求返回的时间，真机调试时，有时需要等待比较长的时间才能返回识别结果。用户也可以直接在微信小程序端发送 HTTP 请求。本书源代码包含两个版本：一个版本是通过云函数发送 HTTP 请求；另一个版本是直接在微信小程序端发送 HTTP 请求。

第 14 章

在微信小程序中使用 ECharts

14.1 在项目中引入 ECharts

ECharts(https://echarts.apache.org/zh/index.html)是一个使用 JavaScript 实现的开源可视化库,可以流畅地运行在 PC 和移动设备上,兼容当前绝大部分浏览器(IE 8/9/10/11、Chrome、Firefox 和 Safari 等),底层依赖矢量图形库 ZRender,提供直观、交互丰富、可高度个性化定制的数据可视化图表。ECharts 提供了常规的折线图、柱状图、散点图、饼图、K 线图,用于统计的盒形图,用于地理数据可视化的地图、热力图、线图,用于关系数据可视化的关系图、树状图、旭日图,多维数据可视化的平行坐标,还有用于 BI 的漏斗图、仪表盘图,并且支持图与图之间的混搭。

因为微信小程序是不支持 DOM 操作的,Canvas 接口也和浏览器不尽相同,所以不能直接使用 ECharts。那么如果在微信小程序中也想使用像 ECharts 这样的可视化工具呢?ECharts 和微信小程序官方团队合作,提供了 ECharts 的微信小程序版本。开发者可以通过熟悉的 ECharts 配置方式,快速开发图表,满足各种可视化需求。

1. 在微信小程序中体验 ECharts

在微信中扫描 ECharts demo 小程序码,即可体验 ECharts 的功能。ECharts demo 小程序码如图 14-1 所示。

图 14-1　ECharts demo 小程序码

2. 下载 ECharts 微信小程序

ECharts 和微信小程序官方团队合作提供了 ECharts 的微信小程序版本，下载 GitHub 上的 ecomfe/echarts-for-weixin 项目（https://github.com/ecomfe/echarts-for-weixin），然后使用微信开发者工具，把 echarts-for-weixin-master 项目导入到工具中，如图 14-2 所示。开发者可以直接将 echarts-for-weixin 项目完全替换新建的项目，然后修改代码作为自己的项目。如果采用完全替换的方式，需要将 project.config.json 中的 AppID 替换成在公众平台申请的项目 ID。pages 目录下的每个文件夹是一个页面，可以根据情况删除不需要的页面，并且在 app.json 中删除对应页面。echarts-for-weixin 项目中 ec-canvas 包含了提供的组件、其他文件是如何使用该组件的示例。ec-canvas 目录下有一个 echarts.js 文件，如果开发者仅使用了 ECharts 中的部分组件同时又想减少文件的大小，可以自行从官方网站自定义构建以减小文件大小。

图 14-2　echarts-for-weixin-master 效果

3. 在自己的项目中引入 ECharts

复制 echarts-for-weixin 项目中的 ec-canvas 目录到自己的微信小程序 miniprogram 目录中，以 index 页面为例，使用 echarts-for-weixin 需要进行的配置如下。

index.json 配置如下：

```
{
  "usingComponents": {
    "ec-canvas": "../../ec-canvas/ec-canvas"
  }
}
```

这一配置的作用是，允许在 index.wxml 中使用< ec-canvas >组件。注意，路径的相对位置要写正确，如果目录结构和本例相同，就应该像上面这样配置。

在 index.wxml 中，创建了一个< ec-canvas >组件，内容如下：

```
1  < view class = "container">
2    < ec - canvas id = "mychart - dom - bar" canvas - id = "mychart - bar" ec = "{{ ec }}"></ ec - canvas >
3  </ view >
```

由于使用了 container 样式，因此需要配置 index.wxss 如下：

```
1  ec - canvas {
2    width: 100%;
3    height: 100%;
4  }
5  .container {
6    position: absolute;
7    top: 180rpx;
8    bottom: 0;
9    left: 0;
10   right: 0;
11   display: flex;
12   flex - direction: column;
13   align - items: center;
14   justify - content: space - between;
15   box - sizing: border - box;
16 }
```

由于在微信小程序中使用了 ColorUI 的导航栏，因此 .container 样式的 top 设置为 180rpx。如果不配置 index.wxss，那么 ECharts 图表将无法在页面上显示。如果多个页面要使用这个样式，可以把样式写入 app.wxss 文件中。

其中，ec 是一个在 index.js 中定义的对象，它使得图表能够在页面加载后被初始化并设置。index.js 的结构如下：

```
1  function initChart(canvas, width, height) {
2    const chart = echarts.init(canvas, null, {
3      width: width,
4      height: height
5    });
6    canvas.setChart(chart);
7
8    var option = {
9      ...
10   };
11   chart.setOption(option);
12   return chart;
13 }
14
15 Page({
```

```
16      data: {
17        ec: {
18          onInit: initChart
19        }
20      }
21    });
```

这对于所有 ECharts 图表都是通用的,用户只需要修改上面 option 的内容,即可改变图表。

14.2 项目主页

本项目挑选了常见的几个 ECharts 图表进行展示,包括柱状图、散点图、折线图以及在一个页面中放多个图表。

项目主页 home.wxml 的代码如下:

```
1   <cu-custom bgColor = "bg-gradual-pink" isBack = "{{false}}">
2       <view slot = "backText">返回</view>
3       <view slot = "content">ECharts</view>
4   </cu-custom>
5
6   <view class = "cu-bar bg-white solid-bottom">
7     <view class = "action">
8       <text class = "cuIcon-title text-orange "></text>应用列表
9     </view>
10  </view>
11  <view class = "cu-list grid col-3 ">
12    <view class = "cu-item" wx:for = "{{boxlist}}" wx:key = "index" wx:for-item = "item">
13      <view data-id = '{{item.id}}' bindtap = 'enterapplication'>
14        <image style = "width: 40px; height: 40px" src = "../../images/{{item.id}}.png"></image>
15        <text>{{item.name}}</text>
16      </view>
17    </view>
18  </view>
```

代码第 1~4 行对应项目头部导航栏。代码第 6~18 行采用 ColorUI 中的宫格列表样式。ECharts 微信小程序主页效果如图 14-3 所示。

相应的 home.js 代码如下:

```
1   Page({
2     data: {
3       boxlist:
4       [
5         {
6           id: 'bar',
7           name: "柱状图",
```

```
8              },
9              {
10                 id: 'scatter',
11                 name:"散点图",
12             },
13             {
14                 id: 'line',
15                 name:"折线图",
16             },
17             {
18                 id: 'multiCharts',
19                 name:"多个图表",
20             }
21         ],
22     },
23     enterapplication: function (event) {
24         var id = event.currentTarget.dataset.id
25         wx.navigateTo({
26             url: '../' + id + '/' + id
27         });
28     }
29 })
```

图 14-3　ECharts 微信小程序主页效果

代码第 3~21 行中 boxlist 定义了含有 4 个元素的数组，对应图 14-3 中的 4 个图表。代码第 23~28 行对应单击不同的图表进入不同的页面，当单击柱状图时，id=bar，这时 url 为

"../bar/bar"；当单击散点图时，id＝scatter，这时 url 为"../scatter/scatter"。

14.3　柱状图页面

柱状图是最常用的图表，柱状图页面 bar.wxml 的代码如下：

```
1   <cu-custom bgColor="bg-gradual-pink" isBack="{{true}}">
2     <view slot="backText">返回</view>
3     <view slot="content">柱状图</view>
4   </cu-custom>
5
6   <view class="container">
7     <button bindtap='updateData'>更新数据</button>
8     <ec-canvas id="mychart-dom-bar" canvas-id="mychart-bar" ec="{{ec}}" bind:init="echartInit"></ec-canvas>
9
10  </view>
```

代码第 6～10 行引入了 <ec-canvas> 组件，ec 中的值为"ec"，因此必须在 bar.js 中 data 中定义一个 ec 的对象。只要使用了 EChart 图表的页面，就必须在相应的 JSON 文件中加入 "usingComponents":{"ec-canvas":"../../ec-canvas/ec-canvas"}，接下来的页面不再重复说明。

bar.wxml 的样式如下：

```
1   ec-canvas {
2     position: absolute;
3     top: 80px;
4     bottom: 0;
5     left: 0;
6     right: 0;
7   }
8   button {
9     margin: 10px;
10  }
11  .container {
12    position: absolute;
13    top: 150rpx;
14    bottom: 0;
15    left: 0;
16    right: 0;
17    display: flex;
18    flex-direction: column;
19    align-items: center;
20    justify-content: space-between;
21    box-sizing: border-box;
22  }
```

柱状图页面显示效果如图 14-4 所示。

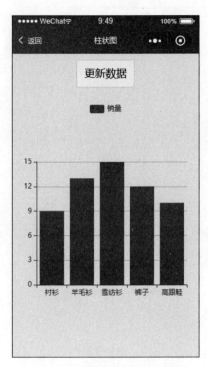

图 14-4 柱状图页面显示效果

相应的 bar.js 代码如下：

```
1   import * as echarts from '../../ec-canvas/echarts';
2   var chart = null
3   function initChart(canvas, width, height) {
4     chart = echarts.init(canvas, null, {
5       width: width,
6       height: height
7     });
8     canvas.setChart(chart);
9     var data = [];
10    for (var i = 0; i < 5; i++) {
11      data.push(
12        Math.round(Math.random() * 20 + 5),
13      )
14    }
15    var option = {
16      tooltip: {},
17      legend: {
18        data: ['销量']
19      },
20      grid: {
21        left: 20,
22        right: 20,
23        bottom: 100,
```

```
24        top: 100,
25        containLabel: true
26      },
27      xAxis: {
28        data: ["衬衫","羊毛衫","雪纺衫","裤子","高跟鞋"]
29      },
30      yAxis: {},
31      series: [{
32        name: '销量',
33        type: 'bar',
34        data: data
35      }]
36    };
37    chart.setOption(option);
38    return chart;
39  }
40  
41  Page({
42    data: {
43      ec: {}
44    },
45    echartInit(e) {
46      initChart(e.detail.canvas, e.detail.width, e.detail.height);
47      var option = chart.getOption();
48      option.series[0].data = [5,10,20,10,20]
49      chart.setOption(option);
50    },
51    updateData: function (event) {
52      var option = chart.getOption();
53      var data = [];
54      for (var i = 0; i < 5; i++) {
55        data.push(
56          Math.round(Math.random() * 20 + 5),
57        )
58      }
59      option.series[0].data = data
60      chart.setOption(option);
61    }
62  })
```

代码第2～39行初始化ECharts图表，其中第4～8行初始化图表，并设置图表的宽度和高度；第9～14行演示图表初始化时数据可以动态生成，这里采用随机函数Math.random，生成含有5个元素的data数组，取值范围为[5,25]；第15～39行设置option选项，option选项的legend.data表示图例的数据数组。数组项通常为一个字符串，每一项都代表一个系列的name；series.type决定了采用的图表类型（每个系列都通过type决定自己的图表类型）；xAxis配置直角坐标系grid中的x轴；yAxis配置直角坐标系grid中的y轴；series.data表示系列中的数据内容数组，第37行把option选项应用到图表中。第42～

44 行 data 中的对象 ec 必须和 wxml 中 ec-canvas 的 ec 属性值相同。第 45~50 行 echartInit 调用 initChart 函数初始化图表,其中第 47~49 行演示如何更改图表数据。第 51~61 行演示单击"更新数据"按钮,更新图表数据,每次单击按钮图表中的数据都会进行更新。需要说明的是,如果希望在页面首次加载时从数据库中读取图表数据,读取数据库的代码可以替换第 9~14 行中的方法,或者替换第 47~49 行中的方法。如果页面加载后希望在某一个事件中动态更新图表中的数据,可以替换第 52~60 行中的方法。

14.4 散点图页面

散点图页面 scatter.wxml 的代码如下:

```
1  <cu-custom bgColor="bg-gradual-pink" isBack="{{true}}">
2      <view slot="backText">返回</view>
3      <view slot="content">散点图</view>
4  </cu-custom>
5
6  <view class="container">
7      <ec-canvas id="mychart-dom-scatter" canvas-id="mychart-scatter" ec="{{ ec }}"></ec-canvas>
8  </view>
```

散点图 scatter.wxss 的代码如下:

```
1  ec-canvas {
2      width: 100%;
3      height: 100%;
4  }
5  .container {
6      position: absolute;
7      top: 150rpx;
8      bottom: 0;
9      left: 0;
10     right: 0;
11     display: flex;
12     flex-direction: column;
13     align-items: center;
14     justify-content: space-between;
15     box-sizing: border-box;
16 }
```

使用 ECharts 图表的页面的样式都类似,因此接下来的页面没有特别说明,其他用到 ECharts 图表的页面样式和 scatter.wxss 一样。散点图页面效果如图 14-5 所示。

相应的 scatter.js 的代码如下:

```
1  import * as echarts from '../../ec-canvas/echarts';
2  function initChart(canvas, width, height) {
3      const chart = echarts.init(canvas, null, {
```

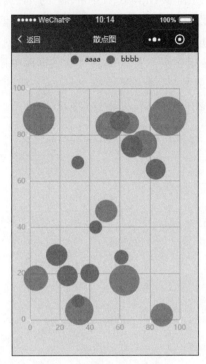

图 14-5 散点图页面效果

```
4        width: width,
5        height: height
6      });
7      canvas.setChart(chart);
8      var data = [];
9      var data2 = [];
10     for (var i = 0; i < 10; i++) {
11       data.push(
12         [
13           Math.round(Math.random() * 100),
14           Math.round(Math.random() * 100),
15           Math.round(Math.random() * 40)
16         ]
17       );
18       data2.push(
19         [
20           Math.round(Math.random() * 100),
21           Math.round(Math.random() * 100),
22           Math.round(Math.random() * 100)
23         ]
24       );
25     }
26     var axisCommon = {
27       axisLabel: {
```

```
      textStyle: {
        color: '#C8C8C8'
      }
    },
    axisTick: {
      lineStyle: {
        color: '#fff'
      }
    },
    axisLine: {
      lineStyle: {
        color: '#C8C8C8'
      }
    },
    splitLine: {
      lineStyle: {
        color: '#C8C8C8',
        type: 'solid'
      }
    }
  };
  var option = {
    color: ["#FF7070", "#60B6E3"],
    backgroundColor: '#eee',
    xAxis: axisCommon,
    yAxis: axisCommon,
    legend: {
      data: ['aaaa', 'bbbb']
    },
    visualMap: {
      show: false,
      max: 100,
      inRange: {
        symbolSize: [20, 70]
      }
    },
    series: [{
      type: 'scatter',
      name: 'aaaa',
      data: data
    },
    {
      name: 'bbbb',
      type: 'scatter',
      data: data2
    }
    ],
    animationDelay: function (idx) {
      return idx * 50;
```

```
77        },
78        animationEasing: 'elasticOut'
79      };
80      chart.setOption(option);
81      return chart;
82    }
83    
84    Page({
85      data: {
86        ec: {
87          onInit: initChart
88        }
89      },
90    });
```

代码第 8～24 行随机生成分别含 10 个随机数的 data 和 data2 数组。第 49～79 行配置图表，series.type 为 scatter，配置图标为散点图；series 包含两个对象数组，配置图表中将显示两组数据；visualMap 是视觉映射组件，用于进行视觉编码，也就是将数据映射到视觉元素（视觉通道）。第 86～88 行对应 ec 图标组件初始化，这里采用了和柱状图不同的初始化方法，柱状图中定义了 echartInit(e) 事件，两种方式效果一样。

14.5 折线图页面

折线图 line.wxml 代码如下：

```
1  <cu-custom bgColor = "bg-gradual-pink" isBack = "{{true}}">
2      <view slot = "backText">返回</view>
3      <view slot = "content">折线图</view>
4  </cu-custom>
5  
6  <view class = "container">
7      <ec-canvas id = "mychart-dom-line" canvas-id = "mychart-line" ec = "{{ ec }}"></ec-canvas>
8  </view>
```

折线图 line.wxss 的样式和 scatter.wxss 的样式一样，折线图页面效果如图 14-6 所示。相应的 line.js 代码如下：

```
1  import * as echarts from '../../ec-canvas/echarts';
2  const app = getApp();
3  function getBarOption() {
4    return {
5      title: {
6        text: '折线图',
7        left: 'center'
8      },
9      color: ["#8B008B", "#FFA500", "#FF0000"],
```

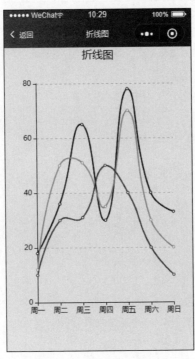

图 14-6　折线图页面效果

```
10      grid: {
11          containLabel: true
12      },
13      tooltip: {
14          show: true,
15          trigger: 'axis'
16      },
17      xAxis: {
18          type: 'category',
19          boundaryGap: false,
20          data: ['周一', '周二', '周三', '周四', '周五', '周六', '周日'],
21      },
22      yAxis: {
23          x: 'center',
24          type: 'value',
25          splitLine: {
26              lineStyle: {
27                  type: 'dashed'
28              }
29          }
30      },
31      series: [{
32          name: 'A',
33          type: 'line',
34          smooth: true,
```

```
35          data: [18, 36, 65, 30, 78, 40, 33]
36        }, {
37          name: 'B',
38          type: 'line',
39          smooth: true,
40          data: [12, 50, 51, 35, 70, 30, 20]
41        }, {
42          name: 'C',
43          type: 'line',
44          smooth: true,
45          data: [10, 30, 31, 50, 40, 20, 10]
46        }]
47      };
48   }
49
50   Page({
51     data: {
52       ec: {
53         onInit: function(canvas, width, height) {
54           const barChart = echarts.init(canvas, null, {
55             width: width,
56             height: height
57           });
58           canvas.setChart(barChart);
59           barChart.setOption(getBarOption());
60           return barChart;
61         }
62       }
63     },
64   })
```

代码第 3~48 行配置图表 option，其中 title 配置组件标题；color 配置折线的颜色；tooltip 配置用户单击图表时显示提示信息；xAxis 配置直角坐标系 grid 中的 x 轴；yAxis 配置直角坐标系 grid 中的 y 轴；series.data 配置三条折线的数据；series.type 为 line，表示为折线图。代码第 52~62 行初始化图表组件，注意和 scatter 初始化图表方式写法上的区别。

14.6 多个图表页面

有时需要在同一个页面上显示多个图表组件，本案例演示一个页面同时显示饼图和仪表盘。多个图表页面 multiCharts.wxml 的代码如下：

```
1  <cu-custom bgColor="bg-gradual-pink" isBack="{{true}}">
2    <view slot="backText">返回</view>
3    <view slot="content">多个图表</view>
4  </cu-custom>
```

```
5
6   <view class = "container">
7     <ec - canvas id = "mychart - dom - multi - pie" canvas - id = "mychart - pie" ec = "{{ ecPie }}">
8   </ec - canvas>
9     <ec - canvas id = "mychart - dom - multi - gauge" canvas - id = "mychart - gauge"
10  ec = "{{ ecGauge }}"></ec - canvas>
11  </view>
```

页面中放了 2 个< ec-canvas >组件，ec 中的值分别为"ecPie"和"ecGauge"。因为页面中放置了 2 个图表组件，所以需要在 multiCharts.wxss 中设置 2 个图表组件的尺寸。multiCharts.wxss 代码如下：

```
1   #mychart - dom - multi - pie {
2     position: absolute;
3     top: 150rpx;
4     bottom: 45%;
5     left: 0;
6     right: 0;
7   }
8   #mychart - dom - multi - gauge {
9     position: absolute;
10    top: 55%;
11    bottom: 0;
12    left: 0;
13    right: 0;
14  }
```

页面中放了 2 个< ec-canvas >组件，多个图表页面效果如图 14-7 所示。

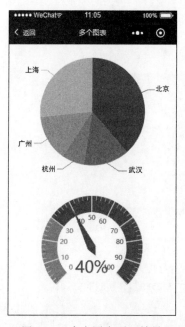

图 14-7　多个图表页面效果

相应的 multiCharts.js 代码如下：

```js
import * as echarts from '../../ec-canvas/echarts';
function getPieOption() {
  return {
    backgroundColor: "#ffffff",
    color: ["#37A2DA", "#32C5E9", "#67E0E3", "#91F2DE", "#FFDB5C", "#FF9F7F"],
    series: [{
      label: {
        normal: {
          fontSize: 14
        }
      },
      type: 'pie',
      center: ['50%', '50%'],
      radius: [0, '80%'],
      data: [{
        value: 55,
        name: '北京'
      }, {
        value: 20,
        name: '武汉'
      }, {
        value: 10,
        name: '杭州'
      }, {
        value: 20,
        name: '广州'
      }, {
        value: 38,
        name: '上海'
      },
      ],
      itemStyle: {
        emphasis: {
          shadowBlur: 10,
          shadowOffsetX: 0,
          shadowColor: 'rgba(0, 2, 2, 0.3)'
        }
      }
    }]
  };
}

function getGaugeOption() {
  return {
    backgroundColor: "#ffffff",
    color: ["#37A2DA", "#32C5E9", "#67E0E3"],
    series: [{
      name: '业务指标',
```

```javascript
49          type: 'gauge',
50          detail: {
51            formatter: '{value}%'
52          },
53          axisLine: {
54            show: true,
55            lineStyle: {
56              width: 30,
57              shadowBlur: 0,
58              color: [
59                [0.3, '#67e0e3'],
60                [0.7, '#37a2da'],
61                [1, '#fd666d']
62              ]
63            }
64          },
65          data: [{
66            value: 40,
67          }]
68
69        }]
70      };
71  }
72
73  Page({
74    data: {
75      ecPie: {
76        onInit: function (canvas, width, height) {
77          const barChart = echarts.init(canvas, null, {
78            width: width,
79            height: height
80          });
81          canvas.setChart(barChart);
82          barChart.setOption(getPieOption());
83
84          return barChart;
85        }
86      },
87      ecGauge: {
88        onInit: function (canvas, width, height) {
89          const scatterChart = echarts.init(canvas, null, {
90            width: width,
91            height: height
92          });
93          canvas.setChart(scatterChart);
94          scatterChart.setOption(getGaugeOption());
95          return scatterChart;
96        }
97      }
98    },
99  });
```

代码第 2～41 行配置饼图 option，其中 backgroundColor 设置了背景颜色；color 设置了饼图中每块的颜色；series.type 设置为 pie，表示为饼图；series.data 设置了饼图中每块的大小和名称。代码第 43～71 行配置仪表盘 option，其中 series.type 设置为 gauge，表示为仪表盘；series.data 设置了当前仪表盘指向的值。代码第 75～97 行分别初始化饼图和仪表盘，初始化的方式和折线图的方式一样。

第 15 章

通过 HTTP API 访问云开发资源

很多时候需要采用网页的形式来管理云数据库和云存储。比如开发了一款新闻类微信小程序，管理员很多时候需要对新闻内容进行编辑，而采用微信小程序对新闻内容进行编辑操作难度大。目前访问云开发资源包含 SDK 调用和使用 HTTP 触发进行调用。其中，SDK 调用提供客户端和服务端支持，客户端 SDK 现阶段支持微信小程序 SDK、JavaScript SDK 和 Flutter SDK；服务端 SDK 现阶段支持 Node.js SDK 和 PHP SDK。本章以使用 HTTP 触发进行调用为例，演示如何访问云开发资源。

微信官方网站开放了 HTTP API 接口，HTTP API 提供了微信小程序外访问云开发资源的能力，使用 HTTP API 开发者可在已有服务器上访问云资源，实现与云开发的互通。本章介绍用 Node.js 访问云开发平台上的云资源，并提供 Web API 接口，读者可以在此基础上，使用 Vue.js 调用这些 Web API 接口来开发相应的网页。

15.1 搭建 Node.js 网站

读者可以采用 Python 提供的 Django 框架、Java 提供的 Spring Boot 框架等开发相应的后端，因为云函数采用的就是 Node.js 语言，所以本节采用 Node.js+Express 搭建 Web 服务器、采用 Axios 请求第三方 HTTP API 访问云开发平台上的云资源，并提供 Web API 接口，读者可以自由选择后端的编程语言，因为访问云开发平台上的云资源实际上就是使用 HTTP 请求对云资源进行访问。

(1) 安装 Node.js，读者自行完成。

(2) 安装 express 模块，运行如下命令：

```
npm install -g express-generator
```

安装完成后读者可以到控制台中输入 express-version，查看能否输出 Express 的版本信息，如果可以看到则表示安装成功；如果不能看到版本信息就需要查看是否正确配置了环境变量。

(3) 生成项目。

打开控制台,进入需要生成项目的路径;在控制台中输入 express myapp,这里 myapp 为项目名称,使用 express 生成项目文件如图 15-1 所示,且在桌面上生成 node_wx 项目文件。

图 15-1 使用 express 生成项目文件

在控制台中输入如下命令:

```
cd node_wx
npm install
npm start
```

其中,cd node_wx 表示进入新创建的 node_wx 目录,npm install 安装相应的依赖库,npm start 开启服务。如果没有错误,在浏览器中输入网址 http://localhost:3000/,就可以看到 Node.js 网站欢迎页,如图 15-2 所示。

图 15-2 Node.js 网站欢迎页

(4) 安装 WebStorm 开发工具。

网页开发可以采用 Atom、VS Code、WebStorm 等,这里使用 WebStorm,安装过程这里不再详细介绍。

开发过程中还用到 axios 发送 HTTP 请求,body-parser 用来解析 POST 的请求,读者可以在 WebStorm 中使用 npm install 命令安装依赖库。

读者可以在 WebStorm 中单击 Run app.js 按钮来运行项目,开启项目后在浏览器中可以进行浏览。

本章主要演示访问云平台上云资源，并提供 Web API 接口，因此对 app.js 进行了简单的配置，配置文件如下：

```
1   var createError = require('http-errors');
2   var express = require('express');
3   var path = require('path');
4   var ejs = require('ejs');
5   var cookieParser = require('cookie-parser');
6   var logger = require('morgan');
7   var bodyParser = require('body-parser');
8   var app = express();
9   app.listen(8080,function(){
10    console.log("服务已开启")
11  })
12  // view engine setup
13  app.set('views', path.join(__dirname, 'views'));
14  // app.set('view engine', 'jade');
15  app.engine('html',ejs.__express);
16  app.set('view engine', 'html');
17  app.use(bodyParser.urlencoded({extended: true}))
18  app.use(logger('dev'));
19  app.use(express.json());
20  app.use(express.urlencoded({ extended: false }));
21  app.use(cookieParser());
22  app.use(express.static(path.join(__dirname, 'public')));
23  app.use('/', require('./routes/index'));
24  app.use('/api', require('./routes/api'));
25  app.use('/admin', require('./routes/admin'));
26  // catch 404 and forward to error handler
27  app.use(function(req, res, next) {
28    next(createError(404));
29  });
30  // error handler
31  app.use(function(err, req, res, next) {
32    // set locals, only providing error in development
33    res.locals.message = err.message;
34    res.locals.error = req.app.get('env') === 'development' ? err : {};
35    // render the error page
36    res.status(err.status || 500);
37    res.render('error');
38  });
39  module.exports = app;
```

代码第 9~11 行配置网站访问端口号 8080。第 15、16 行配置静态网页，采用 html 代替 jade 文件。代码第 17 行配置 body-parser，用来解析 POST 的请求。代码第 23~25 行配置了 3 个路由：普通用户路由、Web API 路由和管理员路由。实际开发中，这里仅仅使用了 Web API 路由，因此代码部分主要在 routers/api.js 文件中。

15.2 访问云平台数据库资源

所有云平台资源的访问都需要使用 access_token 值,access_token 是获取微信小程序全局唯一后台接口调用凭据。调用绝大多数后台接口时都需使用 access_token。

15.2.1 取 access_token 值

请求地址:

```
GET https://api.weixin.qq.com/cgi-bin/token?grant_type=client_credential&appid=APPID&secret=APPSECRET
```

请求参数(见表 15-1):

表 15-1 请求参数 1

属性	类型	是否必填	说明
grant_type	string	是	填写 client_credential
appid	string	是	微信小程序唯一凭证,即 AppID,可在"微信公众平台→设置→开发设置"页中获得
secret	string	是	微信小程序唯一凭证密钥,即 AppSecret,获取方式同 appid

读者先可以使用 Postman 工具测试获取 access_token 的值,如图 15-3 所示。

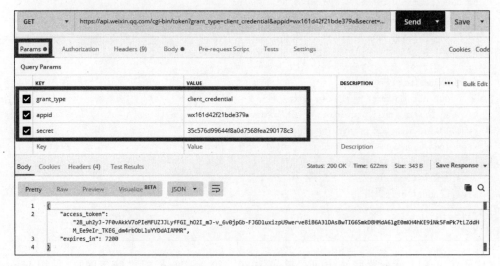

图 15-3 使用 Postman 工具测试获取 access_token 的值

测试成功后,在 api.js 中写获取 access_token 的 Web API 接口,代码如下:

```
1   var express = require('express');
2   var router = express.Router();
3   const axios = require("axios");
4   /* GET http://localhost:8080/api/token. */
5   router.get('/token', function(req, res, next) {
6       let grant_type = "client_credential"
7       let appid = "wx161d42f21bde379a"
8       let secret = "35c576d99644f8a0d7568fea290178c3"
9       let url = "https://api.weixin.qq.com/cgi-bin/token?grant_type=" + grant_type +
10  "&appid=" + appid + "&secret=" + secret
11      axios.get(url)
12          .then(result =>{
13              // handle success
14              console.log(result.data);
15              global.token = result.data.access_token
16              res.json(result.data)
17          })
18          .catch(error =>{
19              // handle error
20              console.log(error);
21          });
22  });
23  module.exports = router;
```

代码第 1～3 行加载相应模块，代码第 5～22 行实现 Web API 接口获取 access_token，使用 Axios 向地址 https://api.weixin.qq.com/cgi-bin/token 发送 GET 请求，读者需要把 appid 和 secret 替换为自己微信小程序的 appid 和 secret，最后使用 res.json 把返回值转换为 JSON 格式的数据进行返回。为了简化，后续的 Web API 接口只写 router 部分。

因为获取 access_token 的 Web API 接口是 GET 请求，可以直接在浏览器中输入网址 http://localhost:8080/api/token 进行调用，也可以在 Postman 中进行调用。调用 Web API 接口获取 access_token 值如图 15-4 所示。

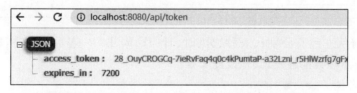

图 15-4　调用 Web API 接口获取 access_token 值

关于 access_token 更新的几点说明：
- access_token 的有效期目前为 2 小时，需定时刷新，但重复获取将导致上次获取的 access_token 失效。
- 建议开发者使用中控服务器统一获取和刷新 access_token，其他业务逻辑服务器所使用的 access_token 均来自该中控服务器，不应该各自去刷新，否则容易造成冲突，导致 access_token 覆盖而影响业务。
- access_token 的有效期通过返回的 expire_in 来传达，目前是 7200 秒内的值，中控服务器需要根据这个有效时间提前去刷新。在刷新过程中，中控服务器可对外继续输出老的 access_token，此时公众平台后台会保证在 5 分钟内新老 access_token 都可用，这保证了第三方业务的平滑过渡。

- access_token 的有效时间可能会在未来有调整，所以中控服务器不仅需要内部定时主动刷新，还需要提供被动刷新 access_token 的接口，这样便于业务服务器在 API 调用获知 access_token 已超时的情况下，可以触发 access_token 的刷新流程。

鉴于 access_token 的更新机制，access_token 的有效期为 2 小时，上面演示了被动刷新 access_token 的接口，而实际上还需要提供内部定时主动刷新，因此需要设置一个定时任务来主动刷新 access_token 的接口，这里使用 node-schedule 来完成定时任务，用户需要在终端安装 node-schedule 依赖库，命令为：npm install node-schedule-save。node-schedule 定时器采用 Cron 风格，使用方法如下：

```
1  const schedule = require('node-schedule');
2  const scheduleCronstyle = () =>{
3      //每分钟的第 30 秒定时执行一次：
4      schedule.scheduleJob('30 * * * * *',() =>{
5          console.log('scheduleCronstyle:' + new Date());
6      });
7  }
8  scheduleCronstyle();
```

schedule.scheduleJob 的回调函数中写入要执行的任务代码，定时器的规则参数如下：

```
        * * * * * *
        │ │ │ │ │ │
        │ │ │ │ │ └ day of week (0 - 7) (0 or 7 is Sun)
        │ │ │ │ └── month (1 - 12)
        │ │ │ └──── day of month (1 - 31)
        │ │ └────── hour (0 - 23)
        │ └──────── minute (0 - 59)
        └────────── second (0 - 59, OPTIONAL)
```

6 个占位符从左到右分别代表秒、分、时、日、月、周几；* 表示通配符，匹配任意，当秒是 * 时，表示任意秒数都触发，其他以此类推。

下面举几个例子。

每分钟的第 30 秒触发：'30 * * * * *'

每小时的 1 分 30 秒触发：'30 1 * * * *'

每天的凌晨 1 点 1 分 30 秒触发：'30 1 1 * * *'

每月的 1 日 1 点 1 分 30 秒触发：'30 1 1 1 * *'

2016 年的 1 月 1 日 1 点 1 分 30 秒触发：'30 1 1 1 2016 *'

每周一的 1 点 1 分 30 秒触发：'30 1 1 * * 1'

在 app.js 中写入定时刷新 access_token，并存入全局变量 global.token 中：

```
1   const schedule = require('node-schedule');
2   const axios = require("axios");
3   function token() {
4     let grant_type = "client_credential";
5     let appid = "wx161d42f21bde379a";
6     let secret = "35c576d99644f8a0d7568fea290178c3";
7     let url = "https://api.weixin.qq.com/cgi-bin/token?grant_type=" + grant_type + "
8   &appid=" + appid + "&secret=" + secret
9     axios.get(url)
10        .then(result =>{
11          // handle success
12          global.token = result.data.access_token
13          console.log(new Date().toLocaleString(),",global token=",global.token);
14        })
15        .catch(error =>{
16          // handle error
17          console.log(error);
18        });
19  }
20  global.token = token()
21  const scheduleCronstyle = () =>{
22    //每分钟的第 30 秒定时执行一次：
23    schedule.scheduleJob('30 1 * * * *',() =>{
24      token()
25    });
26  }
27  scheduleCronstyle();
```

代码第 3~19 行获取 access_token，服务器开启时需要获取一次 access_token，因此第 20 行开启服务器时主动获取 access_token 并存入全局变量 global.token 中，然后开启定时器每小时 1 分 30 秒触发刷新 access_token 值。

15.2.2 数据库导入

请求地址：

POST https://api.weixin.qq.com/tcb/databasemigrateimport?access_token=ACCESS_TOKEN

请求参数(见表 15-2)：

表 15-2　请求参数 2

属　　性	类　　型	是否必填	说　　明
access_token	string	是	接口调用凭证
env	string	是	云环境 ID
collection_name	string	是	导入的集合名称
file_path	string	是	导入文件路径

续表

属性	类型	是否必填	说明
file_type	number	是	导入文件类型
stop_on_error	bool	是	是否在遇到错误时停止导入
conflict_mode	number	是	冲突处理模式

表中，access_token 值的获取前面已经进行了介绍；env 是云环境 ID，打开微信开发者工具，进入云开发控制台，单击右上角的"设置"按钮，在"环境设置"页面中可以看到创建的环境 ID；collection_name 为云数据库中集合名称，用户进行数据导入前，如果没有该集合则需要在云开发控制台中新建数据库集合；file_path 为导入文件的路径，导入文件需先上传到同环境的存储中，可使用开发者工具或 HTTP API 的上传文件 API 上传；file_type 为导入文件的类型，目前仅支持 JSON 和 CSV 两种格式，如果是 JSON 格式的文件，则 file_type 的值为1，如果是 CSV 格式的文件，则 file_type 的值为2；conflict_mode 为冲突处理模式，当数据插入时，出现关键字段重复的数据的处理方式，如果出现重复数据，执行 INSERT（插入）操作，则 conflict_mode 的值为1，如果出现重复数据，执行 UPSERT（更新）操作，则 conflict_mode 的值为2。

接下来演示把一个 JSON 文件导入数据中，首先把 JSON 文件通过开发者工具上传到云存储中，如图 15-5 所示，把本地的 device.json 文件上传到 files 文件夹下，实际开发中需要通过代码把 device.json 文件上传到云存储中。

图 15-5　把 JSON 文件上传到云存储中

因为云数据库中还没有集合，所以需要在数据库中新建集合 device，把云存储 files/device.json 文件导入数据库的 Web API 接口，代码如下：

```
/* POST http://localhost:8080/api/databasemigrateimport */
router.post('/databasemigrateimport',function(req,res){
    console.log(global.token)
    let token = global.token
    axios.post('https://api.weixin.qq.com/tcb/databasemigrateimport?access_token='+token,{
        env:'test-0pmu0',
        collection_name:'device',
        file_path:'files/device.json',
        file_type:1,
        stop_on_error:true,
```

```
11              conflict_mode:1
12          })
13          .then(function (response) {
14              console.log(response.data);
15              res.json(response.data)
16          })
17          .catch(function (error) {
18              console.log(error);
19          })
20  })
```

代码第 4 行获取全局变量 global.token 的值，global.token 的值由定时器定时更新。第 5～19 行向地址 https://api.weixin.qq.com/tcb/databasemigrateimport 发送 POST 请求，env 是云环境 ID，打开微信开发者工具，进入云开发控制台，单击右上角的"设置"按钮，在"环境设置"页面中可以看到创建的环境 ID；collection_name 为云数据库中集合名称；file_path 为导入文件的路径，如图 15-5 所示，file_type 为导入文件的类型，这里导入的是 JOSN 格式，因此 file_type 的值为 1；stop_on_error 选择 true，表示遇到错误时停止导入；conflict_mode 为冲突处理模式，这里选择 INSERT 操作，则 conflict_mode 的值为 1。

因为使用了 POST 请求，这里使用 Postman 来测试 Web API 接口，Postman 测试 http://localhost:8080/api/databasemigrateimport 接口如图 15-6 所示。

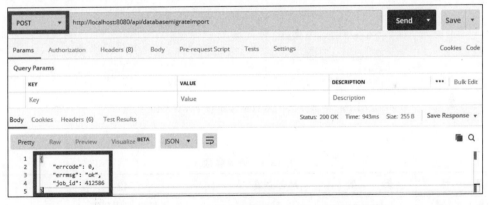

图 15-6 Postman 测试数据库导入 Web API 接口

如果 databasemigrateimport 接口数据导入成功，在 Postman 和网页端都会输出信息：{ errcode：0，errmsg："ok"，job_id：412586 }。数据导入成功后，进入云开发控制台，查看数据库 device 集合，看到成功导入数据，如图 15-7 所示，可以看到 device.json 中有 124 条记录全部导入到了云数据库 device 集合中。

15.2.3 数据库查询记录

请求地址：

POST https://api.weixin.qq.com/tcb/databasequery?access_token = ACCESS_TOKEN

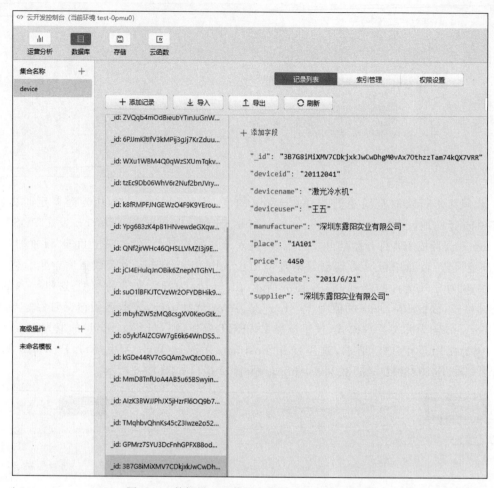

图 15-7　数据导入后数据库集合 device 中的数据

请求参数（见表 15-3）：

表 15-3　请求参数 3

属　　性	类　　型	是否必填	说　　明
access_token	string	是	接口调用凭证
env	string	是	云环境 ID
query	string	是	数据库操作语句

表中，access_token 和 env 值的获取前面已经介绍，query 表示数据库操作语句。用户可以直接使用 Postman 测试 https://api.weixin.qq.com/tcb/databasequery 接口，如图 15-8 所示，因为 query 数据库操作语句包含特色字符，所以这里选择 Body→raw 选项，随后在下拉列表中选择 JSON 选项。

数据库查询记录 Web API 接口代码如下：

```
/* POST http://localhost:8080/api/databasequery */
router.post('/databasequery',function(req,res){
```

```
3       console.log(global.token)
4       let token = global.token
5       axios.post('https://api.weixin.qq.com/tcb/databasequery?access_token = ' + token, {
6           env: 'test-0pmu0',
7           query:'db.collection(\\"device\\").get()'
8       })
9       .then(function (response) {
10          console.log(response.data);
11          res.json(response.data)
12      })
13      .catch(function (error) {
14          console.log(error);
15      })
16  })
```

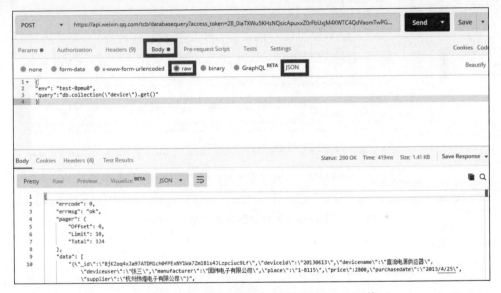

图 15-8　Postman 测试数据库查询记录 Web API 接口

代码第 5～8 行使用 Axios 向地址 https://api.weixin.qq.com/tcb/databasequery 发送 POST 请求，需要注意的是数据库操作语句 query 中有双引号，需要用转移字符\\"来代替引号。

同样采用 Postman 来测试 Web API 接口，Postman 测试 http://localhost:8080/api/databasequery 接口如图 15-9 所示。

从 Postman 的返回结果可以看到 device 集合中总共有 124 条记录，但是每次只能读取 10 条记录，如果需要读取所有数据，需要结合 Collection.skip 指令进行多次读取。用户可以尝试修改 query 数据库操作语句来执行不同的查询操作，比如将 query:'db.collection(\\"device\\").where({price:10000}).get()'的记录查询出来。实际上 query 数据库操作语句可以通过 Web API 接口的 POST 的参数来传递。

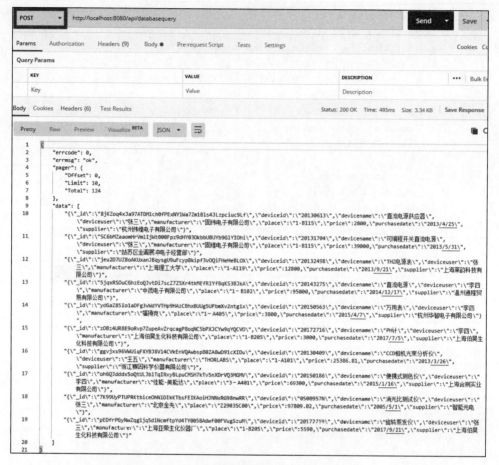

图 15-9　Postman 测试数据库查询 Web API 接口

15.2.4　数据库插入记录

请求地址：

```
POST https://api.weixin.qq.com/tcb/databaseadd?access_token=ACCESS_TOKEN
```

请求参数（见表 15-4）：

表 15-4　请求参数 4

属　　性	类　　型	是否必填	说　　明
access_token	string	是	接口调用凭证
env	string	是	云环境 ID
query	string	是	数据库操作语句

表中，access_token 和 env 值的获取前面已经介绍，query 表示数据库操作语句。

数据库插入记录 Web API 接口代码如下：

```
1   /* POST http://localhost:8080/api/databaseadd */
2   router.post('/databaseadd',function(req,res){
3       let token = global.token
4       axios.post('https://api.weixin.qq.com/tcb/databaseadd?access_token='+token,{
5           env: 'test-0pmu0',
6           query: 'db.collection(\\"contact\\").add({ ' +
7               'data: ' +
8               '[' +
9               '{name:\\"蹇向秋\\",sex:\\"男\\",mobile:\\"15888837968\\"}, ' +
10              '{name:\\"犹淑慧\\",sex:\\"女\\",mobile:\\"13506897988\\"}' +
11              ']})'
12      })
13          .then(function (response) {
14              res.json(response.data)
15          })
16          .catch(function (error) {
17              console.log(error);
18          })
19  })
```

读者需要注意 query 数据库操作语句的写法,使用 https://api.weixin.qq.com/tcb/databaseadd 请求,一次可以向数据库集合中插入多条数据,而微信小程序端和云函数无法通过一次数据库请求插入多条记录。如果云数据库中没有 contact 集合,要先添加 contact 集合。采用 Postman 来测试 Web API 接口,Postman 测试 http://localhost:8080/api/databaseadd 接口如图 15-10 所示。

图 15-10　Postman 测试数据库插入 Web API 接口

从 Postman 返回结果可以看出,2 条记录成功插入数据库。进入云开发控制台,查看数据库 contact 集合,看到成功插入 2 条记录,如图 15-11 所示。

15.2.5　数据库更新记录

请求地址:

```
POST https://api.weixin.qq.com/tcb/databaseupdate?access_token=ACCESS_TOKEN
```

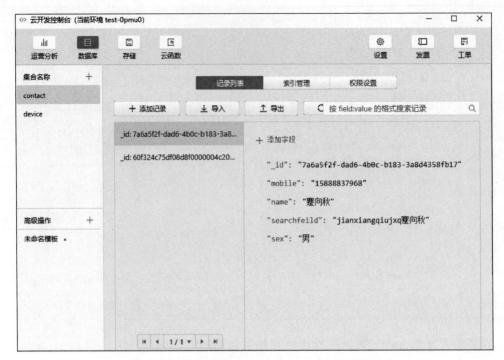

图 15-11　插入数据后数据库集合 contact 中的数据

请求参数（见表 15-5）：

表 15-5　请求参数 5

属　　性	类　　型	是否必填	说　　　明
access_token	string	是	接口调用凭证
env	string	是	云环境 ID
query	string	是	数据库操作语句

表中，access_token 和 env 值的获取前面已经介绍，query 表示数据库操作语句。

数据库更新记录 Web API 接口代码如下：

```
/* POST http://localhost:8080/api/databaseupdate */
router.post('/databaseupdate',function(req,res){
    console.log(global.token)
    let token = global.token
    axios.post('https://api.weixin.qq.com/tcb/databaseupdate?access_token=' + token, {
        env: 'test-0pmu0',
        query: 'db.collection(\\"contact\\").where({name:\\"蹇向秋\\"}).update({data:{sex:\\"女\\"}})'
    })
        .then(function (response) {
            console.log(response.data);
            res.json(response.data)
        })
        .catch(function (error) {
```

```
14                console.log(error);
15            })
16    })
```

代码第 5~8 行使用 Axios 向地址 https://api.weixin.qq.com/tcb/databaseupdate 发送 POST 请求，需要注意的是，数据库操作语句 query 中有双引号，需要用转移字符\\"来代替引号。采用 Postman 来测试 Web API 接口，Postman 测试 http://localhost:8080/api/databaseupdate 接口如图 15-12 所示。进入云开发控制台，查看数据库 contact 集合，name 为"塞向秋"的记录中，性别 sex 修改成了"女"。

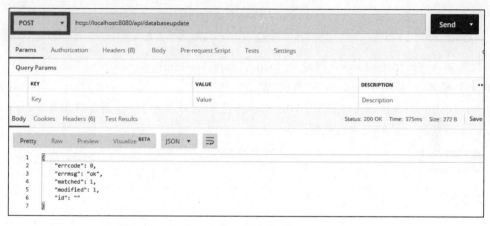

图 15-12 Postman 测试数据库更新 Web API 接口

15.2.6 数据库删除记录

请求地址：

POST https://api.weixin.qq.com/tcb/databasedelete? access_token = ACCESS_TOKEN

请求参数（见表 15-6）：

表 15-6 请求参数 6

属　　性	类　　型	是否必填	说　　明
access_token	string	是	接口调用凭证
env	string	是	云环境 ID
query	string	是	数据库操作语句

表中，access_token 和 env 值的获取前面已经介绍，query 表示数据库操作语句。

数据库删除记录 Web API 接口代码如下：

```
1    /* POST http://localhost:8080/api/databasedelete */
2    router.post('/databasedelete',function(req,res){
3        console.log(global.token)
```

```
4      let token = global.token
5      axios.post('https://api.weixin.qq.com/tcb/databasedelete? access_token = ' + token, {
6          env: 'test-0pmu0',
7          query: 'db.collection(\\"contact\\").where({name:\\"蹇向秋\\"}).remove()'
8      })
9          .then(function (response) {
10             console.log(response.data);
11             res.json(response.data)
12         })
13         .catch(function (error) {
14             console.log(error);
15         })
16     })
```

代码第 5~8 行使用 axios 向地址 https://api.weixin.qq.com/tcb/databasedelete 发送 POST 请求。采用 Postman 来测试 Web API 接口，Postman 测试 http://localhost:8080/api/databasedelete 接口如图 15-13 所示。进入云开发控制台，查看数据库 contact 集合，name 为"蹇向秋"的记录已被删除。

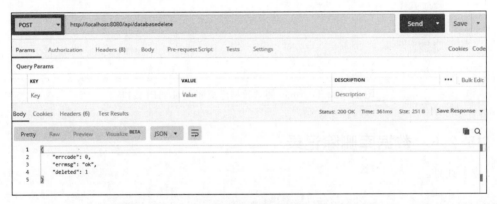

图 15-13　Postman 测试数据库删除记录 Web API 接口

15.3　访问云平台存储资源

15.3.1　获取文件上传链接

请求地址：

```
POST https://api.weixin.qq.com/tcb/batchdownloadfile? access_token = ACCESS_TOKEN
```

请求参数（见表 15-7）：

表 15-7　请求参数 7

属　　性	类　　型	是否必填	说　　明
access_token	string	是	接口调用凭证

续表

属性	类型	是否必填	说明
env	string	是	云环境 ID
path	string	是	上传路径

表中，access_token 获取微信小程序全局唯一后台接口调用凭据，env 是云开发环境 ID，path 表示文件在云存储的路径。

为了演示，这里通过云开发控制台上传 device.json 文件并存储到 files 目录下。单击文件名称，可以查看文件的详细信息，包括文件大小、文件格式、最后更新时间、存储位置、下载地址和 File ID 等，如图 15-14 所示。

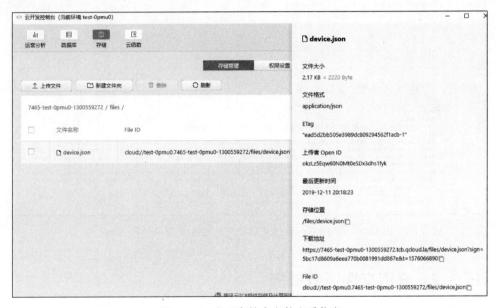

图 15-14 在云存储中文件查看信息

接下来通过 Web API 接口读取文件信息，代码如下：

```
/* POST http://localhost:8080/api/uploadfile */
router.post('/uploadfile',function(req,res){
    console.log(global.token)
    let token = global.token
    axios.post('https://api.weixin.qq.com/tcb/uploadfile?access_token=' + token, {
        env: 'test-0pmu0',
        path: 'files/device.json'
    })
        .then(function (response) {
            console.log(response.data);
            res.json(response.data)
        })
        .catch(function (error) {
            console.log(error);
```

```
15            })
16        })
```

代码第 5～8 行使用 axios 向地址 https://api.weixin.qq.com/tcb/uploadfile 发送 POST 请求。参数 path 表示文件在云存储的路径,如图 15-14 所示。device.json 在云存储中的路径为 files/device.json。采用 Postman 来测试 Web API 接口,Postman 测试 http://localhost:8080/api/uploadfile 接口如图 15-15 所示。从图中可以看出,file_id 字段就是文件上传的链接地址。

图 15-15　Postman 测试获取文件上传链接 Web API 接口

15.3.2　获取文件下载链接

请求地址:

POST https://api.weixin.qq.com/tcb/batchdownloadfile?access_token=ACCESS_TOKEN

请求参数(见表 15-8):

表 15-8　请求参数 8

属　　性	类　　型	是否必填	说　　明
access_token	string	是	接口调用凭证
env	string	是	云环境 ID
file_list	Array < Object >	是	文件列表

表中,access_token 获取微信小程序全局唯一后台接口调用凭据;env 是云开发环境 ID;file_list 表示要查询的文件列表,file_list 是一个对象数组,每个对象包含 fileid(File ID,即文件 ID)和 max_age(下载链接有效期)两个字段。

接下来通过 Web API 接口获取文件下载链接,代码如下:

```javascript
/* POST http://localhost:8080/api/batchdownloadfile */
router.post('/batchdownloadfile',function(req,res){
    console.log(global.token)
    let token = global.token
    axios.post('https://api.weixin.qq.com/tcb/batchdownloadfile?access_token='+token,{
        env: 'test-0pmu0',
        file_list:[
            {fileid:"cloud://test-0pmu0.7465-test-0pmu0-1300559272/files/device.json",max_age:7200},
            {fileid:"cloud://test-0pmu0.7465-test-0pmu0-1300559272/files/设备.xlsx",max_age:7200}
        ]
    })
    .then(function (response) {
        console.log(response.data);
        res.json(response.data)
    })
    .catch(function (error) {
        console.log(error);
    })
})
```

代码第 5～13 行使用 axios 向地址 https://api.weixin.qq.com/tcb/batchdownloadfile 发送 POST 请求。参数 fileid 表示文件在云存储中的 File ID，如图 15-16 所示。该 HTTP API 支持一次查询多个文件的下载链接，本案例查询了 device.json 和设备.xlsx 两个文件的下载链接。采用 Postman 来测试 Web API 接口，Postman 测试 http://localhost:8080/api/batchdownloadfile 接口如图 15-16 所示。从图中可以看出，download_url 字段就是文件上传的链接地址。读者可以把 download_url 字段中的链接地址输入到浏览器中，就可以下载该文件了。

图 15-16　Postman 测试获取文件下载链接 Web API 接口

15.3.3 删除云存储文件

请求地址:

```
POST https://api.weixin.qq.com/tcb/batchdeletefile?access_token = ACCESS_TOKEN
```

请求参数(见表15-9):

表15-9 请求参数9

属　　性	类　　型	是否必填	说　　明
access_token	string	是	接口调用凭证
env	string	是	云环境ID
fileid_list	Array	是	文件ID列表

表中,access_token获取微信小程序全局唯一后台接口调用凭据;env是云开发环境ID;fileid_list表示要删除的文件列表,fileid_list是一个数组,包含要删除的文件ID列表。

接下来通过Web API接口删除云存储文件,代码如下:

```
/* POST http://localhost:8080/api/batchdeletefile */
router.post('/batchdeletefile',function(req,res){
    console.log(global.token)
    let token = global.token
    axios.post('https://api.weixin.qq.com/tcb/batchdeletefile?access_token = '+ token,{
        env: 'test-0pmu0',
        fileid_list:[
            "cloud://test-0pmu0.7465-test-0pmu0-1300559272/files/device.json",
            "cloud://test-0pmu0.7465-test-0pmu0-1300559272/files/设备.xlsx"
        ]
    })
        .then(function (response) {
            console.log(response.data);
            res.json(response.data)
        })
        .catch(function (error) {
            console.log(error);
        })
})
```

代码第5~10行使用axios向地址https://api.weixin.qq.com/tcb/batchdeletefile发送POST请求。参数fileid_list表示要删除的文件在云存储中的File ID,该HTTP API支持一次删除多个文件,采用Postman来测试Web API接口,Postman测试http://localhost:8080/api/batchdeletefile接口如图15-17所示。从Postman返回的值可以看到,delete_list显示了删除文件的情况。

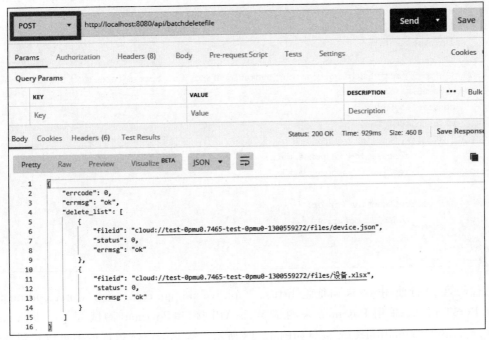

图 15-17　Postman 测试删除存储文件 Web API 接口

15.4　触发云函数

请求地址：

```
POST
https://api.weixin.qq.com/tcb/invokecloudfunction?access_token=ACCESS_TOKEN&env=ENV&name=FUNCTION_NAME
```

请求参数（见表 15-10）：

表 15-10　请求参数 10

属　　性	类　　型	是否必填	说　　明
access_token	string	是	接口调用凭证
env	string	是	云环境 ID
name	string	是	云函数名称
POSTBODY	string	是	云函数的传入参数

表中，access_token 获取微信小程序全局唯一后台接口调用凭据；env 是云开发环境 ID；name 表示云函数名称；POSTBODY 表示云函数的传入参数，具体结构由开发者定义。

接下来通过 Web API 接口触发云函数，这里以触发 5.1 节中发送 HTTP 云函数为例，代码如下：

```
1  /* POST http://localhost:8080/api/invokecloudfunction */
2  router.post('/invokecloudfunction',function(req,res){
```

```
3          console.log(global.token)
4          let token = global.token
5          let name = 'http'
6          let
7   url = 'https://api.weixin.qq.com/tcb/invokecloudfunction? access_token = ' + token + '&env =
8   test - 0pmu0&name = ' + name
9          console.log(url)
10         axios.post(url)
11             .then(function (response) {
12                 console.log(response.data);
13                 res.json(response.data)
14             })
15             .catch(function (error) {
16                 console.log(error);
17             })
18  })
```

代码第 10 行使用 axios 向地址 https://api.weixin.qq.com/tcb/invokecloudfunction 发送 POST 请求。采用 Postman 来测试 Web API 接口，Postman 测试 http://localhost:8080/api/invokecloudfunction 接口如图 15-18 所示。从 Postman 返回的值可以看到，已成功地触发 HTTP 云函数。

图 15-18 Postman 测试触发云函数 Web API 接口

第 16 章

在网页端通过 SDK 访问云开发资源

本章介绍通过 Vue.js 访问云开发资源(云数据库,云存储,云函数)。

16.1 创建 Vue.js 项目

1. 安装 Vue.js

首先检查系统是否安装了 Node 和 NPM,如图 16-1 所示,在控制台中输入 node-v,然后输入 npm-v,就可以查看当前 Node 和 NPM 版本,Node 版本要在 8.9 以上,NPM 版本要在 6 以上;然后在控制台输入如下命令安装 Vue CLI 脚手架:

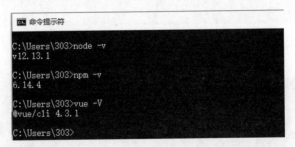

图 16-1　查看 Node、NPM 以及 Vue 版本

```
1  npm install -g @vue/cli
```

其中,-g 表示全局安装 Vue CLI,安装界面如图 16-2 所示。这里需要说明的是,Vue CLI 是用于快速 Vue.js 开发的完整系统,是 Vue.js 开发的官方脚手架。使用脚手架的目的,是其提供了零配置快速原型;提供了运行时依赖;提供了丰富的官方插件库;完整的图形用户界面。

安装完 Vue CLI 脚手架后,可以通过命令行或者软件(Pycharm、HbuilderX、VS Code 和 WebStorm)来创建 Vue.js 项目,这里以使用 WebStorm 为例,演示如何创建 Vue CLI 脚手架项目。

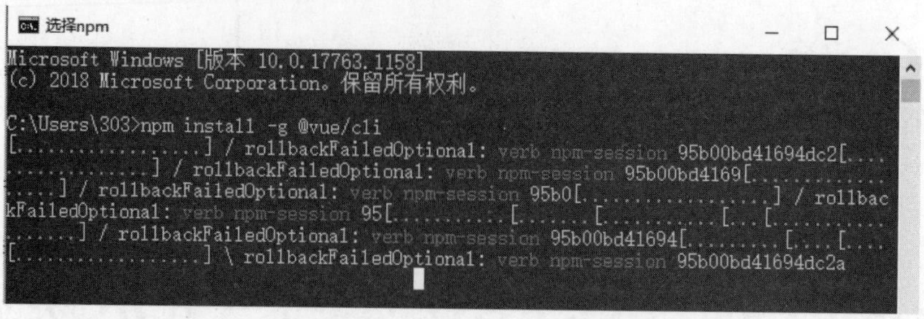

图 16-2　安装 Vue CLI 脚手架界面

打开 WebStorm，在菜单栏中选择 File→New→Project 选项，打开创建新项目界面，如图 16-3 所示，在左侧树形目录中选择 Vue.js，在右侧输入文件存放的位置、Node 和 Vue CLI 脚手架的路径。配置好参数后，单击 Next 按钮，进入下一步，选择 Manually select features 选项并选择需要的配置。勾选 Babel、Router、Vuex、CSS Pre-processors、Linter/Formatter 复选框，如图 16-4 所示。单击 Next 按钮，运行 Vue.js 项目，如图 16-5 所示。

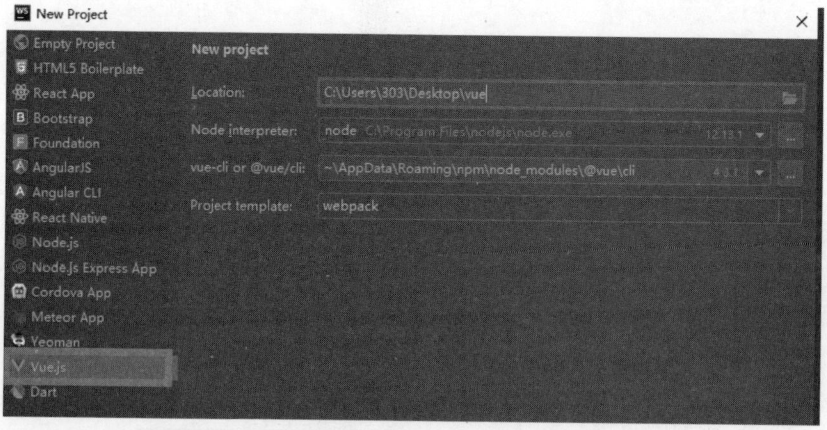

图 16-3　创建 Vue.js 项目

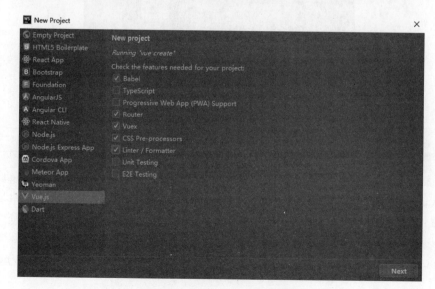

图 16-4　安装 Vue.js 项目配置

第 16 章　在网页端通过SDK访问云开发资源　365

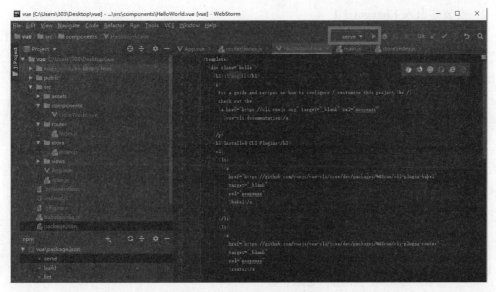

图 16-5　运行 Vue.js 项目

接下来用 WebStrom 打开创建好的项目，在 WebStrom 项目单击"允许"按钮，WebStrom 会对项目进行编译，编译成功会在输出窗口中会显示可以访问的浏览器地址，然后打开浏览器，在浏览器中输入地址 http://localhost:8080/，如果项目运行成功，可以在浏览器中看到 Vue.js 项目主页，如图 16-6 所示。

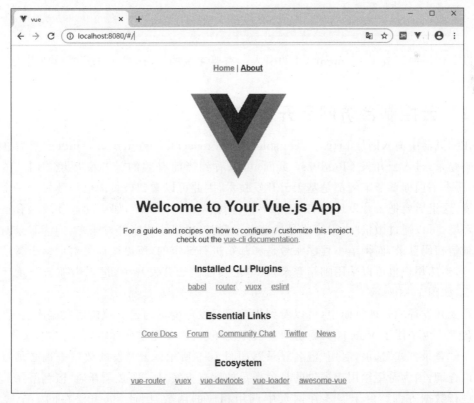

图 16-6　浏览器访问 Vue.js 主页

2. 安装并引入 Element UI

Element UI 是一套为开发者、设计师和产品经理准备的基于 Vue 2.0 的桌面端组件库。推荐使用 NPM 的方式安装 Element UI,它能更好地和 webpack 打包工具配合使用。使用 NPM 安装的命令如下:

```
1  npm i element-ui -S
```

安装完成后可以在创建的 Vue.js 项目中引入 Element UI。在项目中引入 Element UI 分为完整引入 Element UI 和根据需要仅引入部分组件。在实际开发中为了减少打包生成的文件,建议采用根据需要仅引入部分组件。本节为了方便演示,采用完整引入 Element UI。

在 Vue.js 中完整引入 Element UI 需要在 main.js 中写入以下代码:

```
1  import Vue from 'vue';
2  import ElementUI from 'element-ui';
3  import 'element-ui/lib/theme-chalk/index.css';
4  import App from './App.vue';
5
6  Vue.use(ElementUI);
7
8  new Vue({
9    el: '#app',
10   render: h => h(App)
11 });
```

以上代码便完成了 Element UI 的引入。需要注意的是,样式文件需要单独引入。

16.2 云控制台访问云开发资源

在浏览器中输入网址 https://console.cloud.tencent.com/tcb/env/index,然后通过微信扫码登录,进入云开发 CloudBase 页面,可以看到当前有效的云开发环境,如图 16-7 所示,由于本书前面章节案例都是基于云开发技术,因此可以看到该页面已经建好了一个云开发环境,这里需要把云开发环境 ID 记录下来,例如"test-0pmu0",如果读者登录后看不到云开发环境,可以通过微信"小程序账号快速登录"切换微信小程序开发账号(因为登录时采用的是微信扫码登录,而微信小程序账号是和邮箱进行绑定的,所以微信号对应多个微信小程序账号,当然用户也可以采用邮箱登录)。单击当前的云开发环境,进入当前云开发环境管理页面,如图 16-8 所示。

在云开发环境管理页面中,可以看到当前云开发环境下的云资源包括数据库、云存储和云函数等。为了便于 Vue.js 页面访问该云开发环境,需要配置该环境登录方式,选择"环境设置"→"登录方式"选项,启用"匿名登录",如图 16-9 所示。云开发的数据查询必须登录后才可以查询,因为希望给用户提供的是免登录的解决方案,所以,必须开通"匿名登录",确保可以进行数据查询。由于需要在网页中调用相应的函数,因此,也需要在同一个页面的

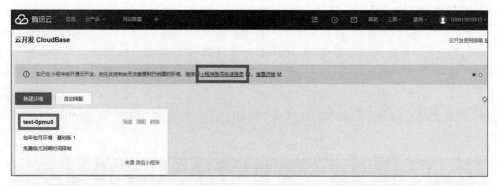

图 16-7　云开发环境

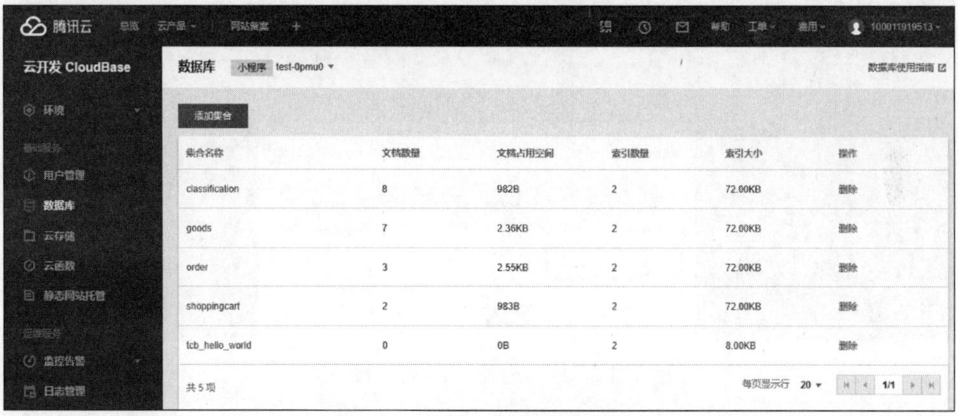

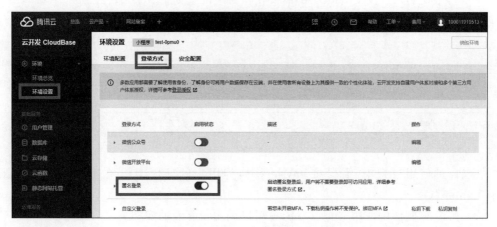

图 16-8　云开发环境管理页面

Web 安全域名中添加应用的上线域名（本地调试用的 localhost 无须添加），本节演示案例只在本地进行调试，因此无须在 Web 安全域名中进行配置。

图 16-9　云开发环境启用"匿名登录"

接下来需要配置 Web 安全域名。在访问云存储的时候，如果不设置安全域名，就无法上传文件到云存储中，项目就会提示错误信息："Access to XMLHttpRequest at 'https://cos.ap-shanghai.myqcloud.com/7465-test-0pmu0-1300559272/timg.jpg' from origin 'http:

//localhost:8080' has been blocked by CORS policy: Response to preflight request doesn't pass access control check: No 'Access-Control-Allow-Origin' header is present on the requested resource."。为了提高文件上传性能,文件上传方式修改为直接上传到对象存储,为了防止在使用过程中出现 CORS 报错,需要到"Web 控制台"→"用户管理"→"登录设置"选项中设置安全域名。如果已有域名出现 CORS 报错,则删除安全域名,重新添加。选择"环境设置"→"安全配置"选项,单击"添加域名"按钮,输入本地域名,例如 localhost:8080,如图 16-10 所示。

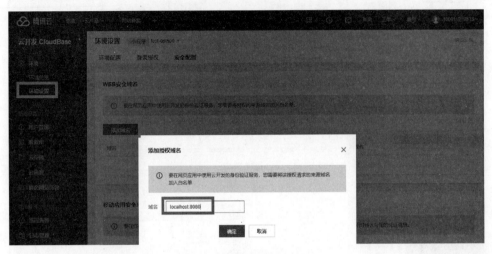

图 16-10　配置 Web 安全域名

16.3　项目主页

在项目中使用云开发 CloudBase,需要在项目中配置如下。
(1) 安装 Vue TCB,在终端中输入命令:

```
1  npm install -- save vue-tcb
```

(2) 在 main.js 中加入如下代码:

```
1  import * as Tcb from 'vue-tcb'
2  Vue.use(Tcb,{env: 'env-id'})
```

在实际的使用过程中,需要将 env-id 替换为用户的环境 ID。在本案例中 env-id 为"test-0pmu0"。
(3) 在 Vue 实例中使用 this.$tcb,就可以调用云开发的实例。
(4) 登录授权,访问云开发资源需要进登录授权,匿名登录命令如下:

```
1  const auth = this.$tcb.auth();
2  auth.signInAnonymously();
```

完整的 main.js 代码如下:

```
1   import Vue from "vue";
2   import App from "./App.vue";
3   import router from "./router";
4   import store from "./store";
5   Vue.config.productionTip = false;
6   import ElementUI from "element-ui";
7   import "element-ui/lib/theme-chalk/index.css";
8   Vue.use(ElementUI);
9   import * as Tcb from "vue-tcb";
10  Vue.use(Tcb, {
11    env: "test-0pmu0"
12  });
13
14  new Vue({
15    router,
16    store,
17    render: h => h(App),
18    beforeCreate: function() {
19      const auth = this.$tcb.auth();
20      auth.signInAnonymously();
21    }
22  }).$mount("#app");
```

代码第 6~8 行完整引入 Element UI 组件库。代码第 9~12 行引入 Vue TCB。代码第 18~21 行对应在生命周期 beforeCreate 中进行匿名登录授权。

项目主页 App.vue 的代码如下:

```
1   <template>
2     <div id="app">
3       <div id="nav">
4         <el-menu
5           :default-active="activeIndex"
6           class="el-menu-demo"
7           mode="horizontal"
8           @select="handleSelect"
9           background-color="#545c64"
10          text-color="#fff"
11          active-text-color="#ffd04b"
12        >
13          <el-menu-item index="1"><a href="#/database">云数据库</a></el-menu-item>
14          <el-menu-item index="2"><a href="#/cloudstorage">云存储</a></el-menu-item>
15          <el-menu-item index="3"><a href="#/cloudfunction">云函数</a></el-menu-item>
16        </el-menu>
17      </div>
18      <router-view />
19    </div>
20  </template>
```

```
21
22    <style lang = "less"></style>
23    <script>
24    export default {
25      data() {
26        return {
27          activeIndex: "1"
28        };
29      },
30      methods: {
31        handleSelect(key, keyPath) {
32          console.log(key, keyPath);
33        }
34      }
35    };
36    </script>
```

代码第 4~16 行采用了 Element UI 中的 NavMenu 导航菜单,导航栏中有 3 个一级菜单,分别对应"云数据库""云存储"和"云函数"。代码第 27 行 activeIndex 变量对应当前选中的菜单。

3 个一级菜单对应 3 个不同的路由,在 router/index.js 路由中配置代码如下:

```
1    import Vue from "vue";
2    import VueRouter from "vue-router";
3    import DataBase from "../components/DataBase.vue";
4    import CloudStorage from "../components/CloudStorage.vue";
5    import CloudFunction from "../components/CloudFunction.vue";
6    Vue.use(VueRouter);
7    const routes = [
8      {path: "/", redirect: '/database'},
9      {path: "/database", component: DataBase},
10     {path: "/cloudstorage", component: CloudStorage},
11     {path: "/cloudfunction", component: CloudFunction},
12   ];
13   const router = new VueRouter({
14     routes
15   });
16   export default router;
```

在路由文件中配置的路由路径"/database"对应 DataBase 组件,路由路径"/cloudstorage"对应 CloudStorage 组件,路由路径"/cloudfunction"对应 CloudFunction 组件。

16.4　云数据库操作

本节演示如何在 Vue.js 中通过 SDK 对云数据库进行增、删、改、查操作,云数据库的详细介绍见第 3 章。

云数据库操作页面 DataBase.vue 的代码如下：

```
1   <template>
2     <div>
3       <el-form :inline="true" :model="classfication" class="demo-form-inline">
4         <el-form-item label="Id">
5           <el-input v-model="classfication.id" placeholder="请输入 id"></el-input>
6         </el-form-item>
7         <el-form-item label="Name">
8           <el-input
9             v-model="classfication.name"
10            placeholder="请输入 Name"
11          ></el-input>
12        </el-form-item>
13        <el-form-item label="Url">
14          <el-input
15            v-model="classfication.url"
16            placeholder="请输入 Url"
17          ></el-input>
18        </el-form-item>
19        <el-form-item>
20          <el-button type="primary" @click="onSubmit">添加记录</el-button>
21        </el-form-item>
22      </el-form>
23
24      <el-table
25        :data="
26          tableData.filter(
27            data =>
28              !search || data.name.toLowerCase().includes(search.toLowerCase())
29          )
30        "
31        style="width: 80%"
32      >
33        <el-table-column label="Id" prop="id" width="200"></el-table-column>
34        <el-table-column label="Name" prop="name" width="200"></el-table-column>
35        <el-table-column label="Url" prop="url" width="300"></el-table-column>
36        <el-table-column align="right">
37          <!-- eslint-disable -->
38          <template slot="header" slot-scope="scope">
39            <el-input v-model="search" size="mini" placeholder="输入关键字搜索" />
40          </template>
41          <template slot-scope="scope">
42            <el-button size="mini" @click="handleEdit(scope.$index)"
43              >Edit</el-button
44            >
45            <el-button
46              size="mini"
47              type="danger"
48              @click="handleDelete(scope.$index, scope.row)"
```

```
49              > Delete </el-button>
50            >
51        </template>
52      </el-table-column>
53    </el-table>
54
55    <!-- Form -->
56    <el-dialog title="商品分类" :visible.sync="dialogFormVisible">
57        <el-form :model="form">
58            <el-form-item label="Id:" :label-width="formLabelWidth">
59                <el-input v-model="form.id" autocomplete="off"></el-input>
60            </el-form-item>
61            <el-form-item label="Name:" :label-width="formLabelWidth">
62                <el-input v-model="form.name" autocomplete="off"></el-input>
63            </el-form-item>
64            <el-form-item label="Url:" :label-width="formLabelWidth">
65                <el-input v-model="form.url" autocomplete="off"></el-input>
66            </el-form-item>
67        </el-form>
68        <div slot="footer" class="dialog-footer">
69            <el-button @click="dialogFormVisible = false">取消</el-button>
70            <el-button type="primary" @click="editSubmit">确定</el-button>
71        </div>
72    </el-dialog>
73  </div>
74 </template>
75 <script>
76 export default {
77   name: "DataBase",
78   data: function() {
79     return {
80       tableData: [],
81       search: "",
82       classfication: {
83         id: "",
84         name: ""
85       },
86       dialogFormVisible: false,
87       form: {
88         _id: "",
89         id: "",
90         name: "",
91         url: ""
92       },
93       formLabelWidth: "80px",
94       selectRow: 0
95     };
96   },
97   methods: {
```

```js
 98      onSubmit() {
 99        console.log("submit!");
100        const db = this.$tcb.database();
101        var classfication = this.classfication;
102
103        db.collection("classification")
104          .add(classfication)
105          .then(res =>{
106            this.searchAll();
107            this.classfication.id = "";
108            this.classfication.name = "";
109            this.classfication.url = "";
110          });
111      },
112      handleEdit(index) {
113        this.dialogFormVisible = true;
114        this.form._id = this.tableData[index]._id;
115        this.form.id = this.tableData[index].id;
116        this.form.name = this.tableData[index].name;
117        this.form.url = this.tableData[index].url;
118        this.selectRow = index;
119        console.log(index);
120      },
121      editSubmit() {
122        const db = this.$tcb.database();
123        var form = this.form;
124        var selectRow = this.selectRow;
125        this.dialogFormVisible = false;
126        this.tableData[selectRow] = form;
127        console.log("form:", form);
128        db.collection("classification")
129          .doc(form._id)
130          .update({
131            id: form.id,
132            name: form.name,
133            url: form.url
134          })
135          .then(res =>{});
136      },
137      handleDelete(index, row) {
138        console.log(index, row);
139        console.log(row._id);
140        var _id = row._id;
141        this.tableData.splice(index, 1);
142        console.log(this.tableData);
143        const db = this.$tcb.database();
144        db.collection("classification")
145          .doc(_id)
146          .remove()
```

```
147         .then(res =>{
148           console.log(res);
149         });
150     },
151     searchAll() {
152       const db = this.$tcb.database();
153       db.collection("classification")
154         .get()
155         .then(res =>{
156           this.tableData = res.data;
157           console.log(this.tableData);
158         });
159     }
160   },
161   mounted() {
162     this.searchAll();
163   }
164 };
165 </script>
166 <style scoped></style>
```

Vue.js 对云数据库的操作页面如图 16-11 所示。代码第 3～22 行对应 Element UI 中的 Form(表单)组件,用于往数据库集合中添加记录。代码第 24～53 行对应 Element UI 中 Table(表格)组件,其中代码第 39 行用于表格记录搜索;代码第 41～44 行对记录进行编辑操作;代码第 45～50 行对记录进行删除操作。代码第 56～72 行对应 Element UI 中 Dialog(对话框)组件。代码第 80 行对应 Table 中的数据。代码第 98～111 行对应"添加记录"按钮,其中代码第 103～110 行往数据库集合 classification 中添加一条记录;this.searchAll()用来查询当前 classification 集合中所有记录。代码第 112～120 行对应 Table

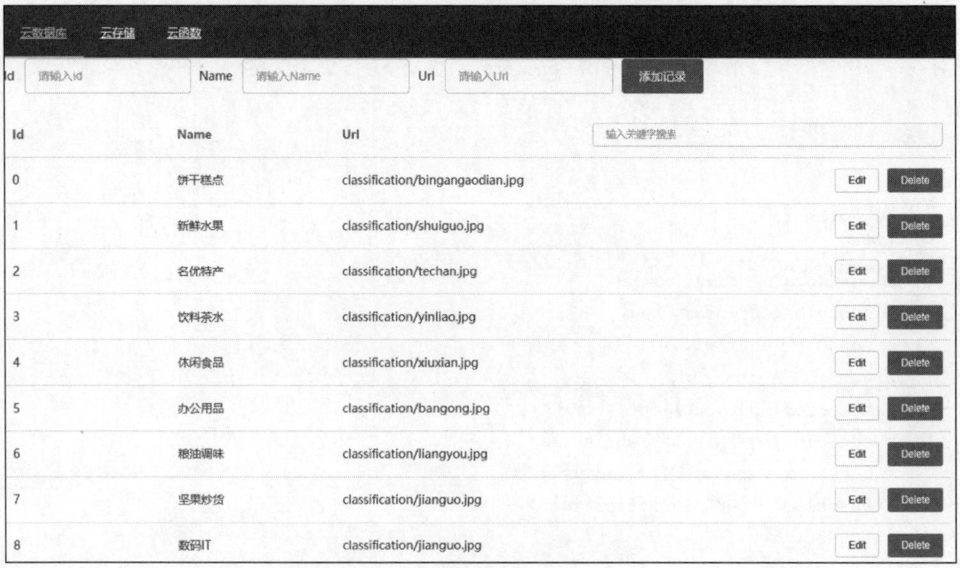

图 16-11　Vue.js 对云数据库的操作页面

中的 Edit 按钮，单击 Edit 按钮，打开 Dialog 对话框，这时需要把 Table 中的记录数据自动填入到 Dialog 对话框中，如图 16-12 所示。此时可以对该记录进行修改，修改完单击"确定"按钮完成数据库记录修改。代码第 121~136 行对应单击"确定"按钮事件，其中第 128~135 行对应数据库单条记录更新操作。代码第 137~150 行对应图 16-11 中的 Delete 按钮，其中第 144~149 行实现云数据库的删除操作。代码第 151~159 行实现查询云数据库集合 classification 中的所有记录。

图 16-12　对记录进行修改页面

16.5　云存储操作

云存储操作包含上传文件、删除文件和下载文件操作。云存储操作页面 CloudStorage.vue 代码如下：

```
1    <template>
2      <div>
3        <el-upload
4          class="upload-demo"
5          drag
6          action="void"
7          :http-request="uploadSectionFile"
8          :on-remove="handleRemove"
9          :on-preview="handlePreview"
10         style="width:400px"
11       >
12         <i class="el-icon-upload"></i>
13         <div class="el-upload__text">将文件拖到此处，或<em>单击上传</em></div>
14         <div class="el-upload__tip" slot="tip">
15           只能上传 jpg/png 文件，且不超过 500KB
```

```html
            </div>
        </el-upload>
    </div>
</template>

<script>
export default {
    name: "Storage",
    data() {
        return {};
    },
    methods: {
        uploadSectionFile: function(params) {
            var file = params.file;
            this.$tcb.uploadFile({
                cloudPath: file.name,
                filePath: file,
                onUploadProgress: progressEvent =>{
                    const percentCompleted = Math.floor(
                        (progressEvent.loaded * 100) / progressEvent.total
                    );
                    params.onProgress({ percent: percentCompleted });
                }
            });
        },
        handleRemove(file, fileList) {
            this.$tcb
                .deleteFile({
                    fileList: [
                        "cloud://test-0pmu0.7465-test-0pmu0-1300559272/" + file.name
                    ]
                })
                .then(res =>{
                    console.log(res);
                });
        },
        handlePreview(file) {
            this.$tcb
                .downloadFile({
                    fileID: "cloud://test-0pmu0.7465-test-0pmu0-1300559272/" + file.name
                })
                .then(res =>{
                    console.log(res);
                });
        }
    }
};
</script>
<style scoped></style>
```

Vue.js 对云存储操作页面如图 16-13 所示,用户单击"单击上传"按钮,从本地选择需要上传的文件。代码第 3~17 行对应 Element UI 中的 Upload(上传)组件,其中,第 7 行 http-request 覆盖默认的上传行为,可以自定义上传的实现;第 8 行 on-remove 对应文件列表移除文件时的钩子;第 9 行 on-preview 对应单击文件列表中已上传的文件时的钩子。第 14~16 行显示提示的说明文字,需要说明的是,本节演示时并没有对上传文件的类型和大小做限制。代码第 28~40 行对应自定义上传的实现,其中第 30~39 行对应上传文件到云存储中,cloudPath 对应文件的绝对路径,包含文件名;filePath 对应要上传的文件对象;onUploadProgress 对应上传进度回调。用户单击文件列表后的"×"按钮,调用 handleRemove 事件。代码第 41~51 行对应删除文件事件,其中第 42~50 行对应从云存储中删除文件,fileList 对应要删除的文件 ID 组成的数组。用户单击文件列表中的文件名,调用 handlePreview 事件,代码第 52~60 行对应下载文件事件,其中第 53~59 行对应从云存储中下载文件,fileID 对应要下载的文件的 ID。图 16-14 显示了上传文件到云存储后,在腾讯云云开发 CloudBase 页面,选择"云存储"→"文件管理"选项后云存储中的文件列表。

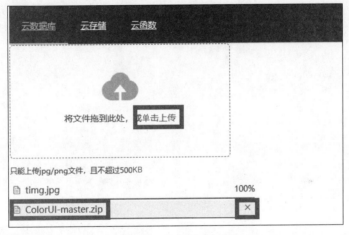

图 16-13　云存储操作页面

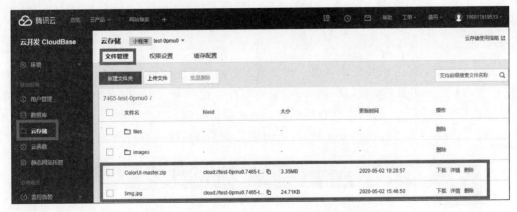

图 16-14　云开发 CloudBase 中云存储中的文件列表

16.6 云函数操作

云函数操作相对比较简单,云函数操作页面 CloudFunction.vue 代码如下:

```
1  <template>
2    <div>
3      <el-button type="primary" @click="onSubmit">触发云函数</el-button>
4      <h3>{{ msg }}</h3>
5    </div>
6  </template>
7
8  <script>
9  export default {
10   name: "CloudFunction",
11   data() {
12     return {
13       msg: ""
14     };
15   },
16   methods: {
17     onSubmit() {
18       console.log("getUser");
19       this.$tcb
20         .callFunction({
21           name: "getUser",
22           data: {}
23         })
24         .then(res => {
25           const result = res.result;       //云函数执行结果
26           console.log(result);
27           this.msg = result;
28         });
29     }
30   }
31 };
32 </script>
33 <style scoped></style>
```

图 16-15 显示了云函数操作页面。代码第 3 行对应 Element UI 中的 Button(按钮)组件,单击"触发云函数"按钮,调用 onSubmit 事件。代码第 17~29 行对应触发云函数 getUser,其中第 19~28 行对应执行云函数,name 字段对应云函数名称,data 字段对应云函数数据。执行触发云函数 getUser 后,在页面中显示了云函数的返回值。

图 16-15 云函数操作页面

图书资源支持

感谢您一直以来对清华版图书的支持和爱护。为了配合本书的使用,本书提供配套的资源,有需求的读者请扫描下方的"书圈"微信公众号二维码,在图书专区下载,也可以拨打电话或发送电子邮件咨询。

如果您在使用本书的过程中遇到了什么问题,或者有相关图书出版计划,也请您发邮件告诉我们,以便我们更好地为您服务。

我们的联系方式:

地　　址:北京市海淀区双清路学研大厦 A 座 701

邮　　编:100084

电　　话:010-83470236　010-83470237

资源下载:http://www.tup.com.cn

客服邮箱:2301891038@qq.com

QQ:2301891038(请写明您的单位和姓名)

资源下载、样书申请

书 圈

扫一扫,获取最新目录

课 程 直 播

用微信扫一扫右边的二维码,即可关注清华大学出版社公众号"书圈"。